Advances in
Electronic Ceramics II

Advances in Electronic Ceramics II

A Collection of Papers Presented at the 33rd International Conference on Advanced Ceramics and Composites January 18–23, 2009 Daytona Beach, Florida

Edited by
Shashank Priya
Anke Weidenkaff
David P. Norton

Volume Editors
Dileep Singh
Jonathan Salem

A John Wiley & Sons, Inc., Publication

Published by John Wiley & Sons, Inc., Hoboken, New Jersey.
Published simultaneously in Canada.

Limit of Liability/Disclaimer of Warranty: While the publisher and author have used their best efforts in
preparing this book, they make no representations or warranties with respect to the accuracy or
completeness of the contents of this book and specifically disclaim any implied warranties of
merchantability or fitness for a particular purpose. No warranty may be created or extended by sales
representatives or written sales materials. The advice and strategies contained herein may not be
suitable for your situation. You should consult with a professional where appropriate. Neither the
publisher nor author shall be liable for any loss of profit or any other commercial damages, including
but not limited to special, incidental, consequential, or other damages.

For general information on our other products and services or for technical support, please contact our
Customer Care Department within the United States at (800) 762-2974, outside the United States at
(317) 572-3993 or fax (317) 572-4002.

Wiley also publishes its books in a variety of electronic formats. Some content that appears in print may
not be available in electronic format. For information about Wiley products, visit our web site at
www.wiley.com.

Library of Congress Cataloging-in-Publication Data is available.

ISBN 978-0-470-45759-7

Printed in the United States of America.

10 9 8 7 6 5 4 3 2 1

Contents

THERMAL TO ELECTRIC CONVERSION

MATERIALS FOR SOLID STATE LIGHTING

Preface

This proceedings issue, *Advances in Electronic Ceramics II,* contains a collection of 12 papers presented during the 33rd International Conference on Advanced Ceramics and Composites, Daytona Beach, FL, January 18–23, 2009. Papers in this proceedings were presented from one of the three symposia listed below:

Symposium 6	Key Materials and Technologies for Efficient Direct Thermal-to-Electrical Conversion
Symposium 11	Symposium on Advanced Dielectrics, Piezoelectric, Ferroelectric, and Multiferroic Materials
Focused Session 2	Materials for Solid State Lighting

The lead paper was presented during the Plenary Session by Dr. Andreas Schön-ecker (IKTS Fraunhofer, Germany) and provides an excellent overview of the correlations between advanced piezoceramics technologies and the development of smart devices, microsystems, and structures.

The editors thank all of the authors who presented a paper or poster during the conference and to the authors who submitted a paper for this proceedings. A special thanks also goes out to all of the organizers and session chairs for their contribution and hard work before, during and after the meeting.

SHASHANK PRIYA
Virginia Polytechnic Institute and State University, USA

ANKE WEIDENKAFF
EMPA, Switzerland

DAVID P. NORTON
University of Florida, USA

Introduction

The theme of international participation continued at the 33rd International Conference on Advanced Ceramics and Composites (ICACC), with over 1000 attendees from 39 countries. China has become a more significant participant in the program with 15 contributed papers and the presentation of the 2009 Engineering Ceramic Division's Bridge Building Award lecture. The 2009 meeting was organized in conjunction with the Electronics Division and the Nuclear and Environmental Technology Division.

Energy related themes were a mainstay, with symposia on nuclear energy, solid oxide fuel cells, materials for thermal-to-electric energy conversion, and thermal barrier coatings participating along with the traditional themes of armor, mechanical properties, and porous ceramics. Newer themes included nano-structured materials, advanced manufacturing, and bioceramics. Once again the conference included topics ranging from ceramic nanomaterials to structural reliability of ceramic components, demonstrating the linkage between materials science developments at the atomic level and macro-level structural applications. Symposium on Nanostructured Materials and Nanocomposites was held in honor of Prof. Koichi Niihara and recognized the significant contributions made by him. The conference was organized into the following symposia and focused sessions:

Symposium 1	Mechanical Behavior and Performance of Ceramics and Composites
Symposium 2	Advanced Ceramic Coatings for Structural, Environmental, and Functional Applications
Symposium 3	6th International Symposium on Solid Oxide Fuel Cells (SOFC): Materials, Science, and Technology
Symposium 4	Armor Ceramics
Symposium 5	Next Generation Bioceramics
Symposium 6	Key Materials and Technologies for Efficient Direct Thermal-to-Electrical Conversion
Symposium 7	3rd International Symposium on Nanostructured Materials and Nanocomposites: In Honor of Professor Koichi Niihara
Symposium 8	3rd International symposium on Advanced Processing & Manufacturing Technologies (APMT) for Structural & Multifunctional Materials and Systems

The conference proceedings compiles peer reviewed papers from the above symposia and focused sessions into 9 issues of the 2009 Ceramic Engineering & Science Proceedings (CESP); Volume 30, Issues 2-10, 2009 as outlined below:

- Mechanical Properties and Performance of Engineering Ceramics and Composites IV, CESP Volume 30, Issue 2 (includes papers from Symp. 1 and FS 1)
- Advanced Ceramic Coatings and Interfaces IV Volume 30, Issue 3 (includes papers from Symp. 2)
- Advances in Solid Oxide Fuel Cells V, CESP Volume 30, Issue 4 (includes papers from Symp. 3)
- Advances in Ceramic Armor V, CESP Volume 30, Issue 5 (includes papers from Symp. 4)
- Advances in Bioceramics and Porous Ceramics II, CESP Volume 30, Issue 6 (includes papers from Symp. 5 and Symp. 9)
- Nanostructured Materials and Nanotechnology III, CESP Volume 30, Issue 7 (includes papers from Symp. 7)
- Advanced Processing and Manufacturing Technologies for Structural and Multifunctional Materials III, CESP Volume 30, Issue 8 (includes papers from Symp. 8)
- Advances in Electronic Ceramics II, CESP Volume 30, Issue 9 (includes papers from Symp. 11, Symp. 6, FS 2 and FS 3)
- Ceramics in Nuclear Applications, CESP Volume 30, Issue 10 (includes papers from Symp. 10 and FS 4)

The organization of the Daytona Beach meeting and the publication of these proceedings were possible thanks to the professional staff of The American Ceramic Society (ACerS) and the tireless dedication of the many members of the ACerS Engineering Ceramics, Nuclear & Environmental Technology and Electronics Divisions. We would especially like to express our sincere thanks to the symposia organizers, session chairs, presenters and conference attendees, for their efforts and enthusiastic participation in the vibrant and cutting-edge conference.

DILEEP SINGH and JONATHAN SALEM
Volume Editors

Dielectric, Piezoelectric, Ferroelectric and Multiferroic Materials

PIEZOELECTRIC COMPOSITE MATERIALS AND STRUCTURES

Andreas J. Schönecker
Fraunhofer Institute Ceramic Technologies and Systems
Winterbergstr.28
D- 01277 Dresden, Germany

ABSTRACT

The present paper considers the correlations between advanced piezoceramic technologies and the development of smart devices, microsystems and structures. Bob Newnham proposed the idea of smart materials in the 70 ties. Since that we find research programs focused on systems with embedded actuators, sensors and controller units. Active structures proved beneficial, which for example arises from increased safety, reduced energy consumption, extended life cycle and unique performance. But as a non-series product they failed often by high production costs. In other words, extension of market share to active structures by commercial applications requires technology chains and designs, which are compatible with mass production. An effective approach to overcome this obstacle might be the use of microsystems / MEMS technology as intermediate step providing pre-integration of active functions.

Technological developments of the recent past following the here expressed strategy are summarized in the paper. We tried to bridge the gap between piezoceramic transducer and smart structure fabrication by qualification and adjustment of high efficient production methods of piezoceramic transducers (extrusion spinning, molding and screen printing), electronic circuits (electronics packaging and electronics production) and structural components (metal die casting). Examples will be given related to integrated thick films in silicon technology, and to ceramic multilayer, polymer and metal matrix architectures.

INTRODUCTION

The present paper considers piezoelectric ceramics as key functional material in composites and structures. Most of piezoelectrics presently exploited commercially are solid solutions based on lead zirconate titanate (PZT) ceramics. Still, compositional developments within the PZT family are performed to meet custom requirements perfectly[1]. Lead free piezoelectrics, such as the sodium potassium niobate solid solution[2,3,4] and bismuth sodium titanate solid solutions[5,6,7] became the topic of much research at the end of the 1990s, which is due to increased environmental awareness[8].

As part of these investigations textured microstructures were studied, which may be interpreted as ceramic composite approach in the microstructure scale. New fabrication processes of piezoelectric ceramics are desirable through which the texture in the ceramic can be well controlled to give preferred grain orientations. A controlled grain structure during the ceramic sintering process gives the ceramics an anisotropic pseudo-single crystal behaviour[1] and may improve key parameters, considerably.

The development of smart structures, as considered in the present paper, involves the integration of piezoceramics and further dissimilar materials performing separate functions into one device. The focus of the present paper is given to commercial PZT ceramics, but the ideas may be extended to new piezoceramics as well. Essentially, the challenge lies in the availability of complete and adjusted manufacturing chains and in the maintenance of the properties of constituent materials during packaging into an individual unit.

Bob Newnham proposed the idea of smart materials in the 70[ties]. Since that we find research programs focused on systems with embedded actuators, sensors and controller units. Active structures proved beneficial, which for example arises from increased safety, reduced energy consumption, extended life cycle and unique performance. But as a non-series product they failed often by high production costs. In other words, extension of market share of active structures to commercial applications requires technology chains and designs, which are compatible with mass production, like typical found in care manufacturing.

An effective approach to bridge the gap between piezoceramic transducer and smart structure fabrication might be the use of microsystems / MEMS technology. It is well established and industrialized and well suited to make use of the potential of active materials in custom devices. It guaranties for pre-integration of sensing, actuation and control. Active structures may then be designed by integration of single or cross-linked piezo-electronic modules. As consequence, integration of these microsystems under the technological conditions of smart structure fabrication, especially in terms of process temperature, mechanical impacts and cycle time, is required.

The present paper summarizes the potential of advanced, microsystems compatible piezo technologies for active structures and systems. Examples will be given related to the integration in silicon wafer, ceramic multilayer, polymer and metal matrix architectures.

KEY CERAMIC TECHNOLOGIES

This chapter considers extrusion spinning, soft molding and tape casting as efficient green forming technologies of single element transducers. High performance piezoceramic units can be obtained and further processed to piezocomposite materials. Screen printing is well suited for the preparation of flextensional transducers that amplifies and changes the direction of generated displacement.

Single element transducer fabrication
PZT fibers by spinning

An essential motivation for the development of piezoceramic fibers arose from the concept of smart fiber-reinforced structural materials with integrated fibers for sensing and actuation [9]. Powder suspension based piezoceramic fibers have been developed to make high performance, single fibers in the thickness range of 100–1000μm available. Suspension extrusion[10] and suspension spinning [11,12,13] were used as basic process routes allowing for different fiber cross sections, like cylinder shaped fibers, hollow fibers and rectangular shaped fibers. See Fig. 1.

Figure 1. Cross sections of PZT fibers. Data are given in Table 1 and the text.

Suspension spinning is considered as well suited for mass production because basically textile fabrication methods are used. Two process variants [11, 13] have been commercialized.

In 1999, Smart Material Corp. (Florida) in co-operation with smartfiber AG, Germany, established the production of PZT fibers by the ALCERU® process[14]. They are made from Type

II and Type VI (U.S. Navy designation standards) piezoceramics and are offered in the diameter range between 105 μm and 1,000 μm. See Table 1. Smart Material Corp. launched their 1–3 piezoelectric fiber composites on the market, allowing for fiber composite engineering and production. A variety of custom products have become available since that.

Table 1: Commercial PZT fibers and tubes offered by Smart Material

Fiber diameters	d : 105 μm ... 1000 μm
Tube outer diameters	d_a : 400 μm and 1000 μm
Tube wall thickness	$d_a - d_i > 100$ μm
Fiber and tube length	< 200 mm
Piezoceramic materials	Type: PZT 4, PZT 5A , PZT 5H

A third process, the polysulfon spinning technology, has recently been developed by Fraunhofer IKTS and TU Dresden for PZT fiber fabrication with the benefit of room temperature processing and the use of environmental harmless solvent NMP (N-methylpyrrolidon) in small quantities[15]. The process chart of the polysulphone process is sketched in Fig. 2.

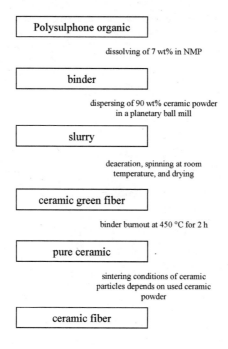

Figure 2. Workflow for the production of ceramic fibers by the polysulphone process[15].

The reduction of porosity and vacuoles turned out to be the main issue. As result of systematically process development sintered fibers with low porosity of < 5 % can be fabricated. The sintered fibers are straight with homogeneous consistency, showing a line fraction of parallel aligned fibers up to 97 %. See Fig. 3. This offers a high attainable volume concentration if processed to composites. Access to all technological steps allows now for processing of fiber with custom key functional data.

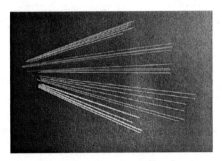

Fig. 3 shows ceramic fibers prepared by the polysulphone process[15]. The Diameter d can be varied between 100 μm … 1000 μm.

PZT plates by tape casting

Basically, PZT plates are prepared by tape casting, drying and sintering under controlled process conditions. As result, single rectangular shaped PZT wafers are obtained. Different contours can be processed by Laser cutting (Fig 4). Dicing is successfully used to process rectangular shaped fibers, already parallel arranged on carrier foils. This is the key ceramic process in MFC fabrication.

Fig. 4: PZT units by tape casting. Left: green tapes, right "tailored" units.

PZT array by soft molding

Arrays of distributed pillars can be manufactured using a patented technology that was invented at the Fraunhofer IKTS, Germany [16,17]. This process consists of copying a soft mold from a positive form of the final structure, filling the mold with a slip loaded with the calcined piezoceramic powder and subsequent firing of the element in PbO-controlled atmosphere.

Master molds can be prepared by micromachining, chemical or plasma etching. For master molds with cylindrical rods, an anisotropic silicon etching (ASET) process is usually applied. From the master molds, numerous templates of a soft plastic can be taken and then filled with the ceramic slurry. Using a soft transfer mold and a ceramic slip with high green strength, the dried ceramic green body can be demolded with low amounts of defects.

This soft-mold process holds many advantages over conventional die-and-fill, injection molding or dicing techniques: Soft molds are reusable. There is no need for expensive cutting machinery. The process allows for free design of elements with various types of fiber shapes, sizes, and pitches, and high volume production while maintaining superior quality at a reasonable price per unit. See Fig. 5.

Figure 5. Ceramic pillar array unit prepared by soft molding.

Piezocomposite transducer fabrication
Piezoceramic thick films by screen printing

Screen printing is a very flexible, cost-effective and scalable technology for producing functional layers on green and fired substrates. The preparation of ferroelectric $Pb(Zr_xTi_{1-x})O_3$ (PZT) thick films on various substrates has been an intensive research field of IKTS with focus on the application as information storage material in electrostatic printing machines, and integrated sensors and actuators. PZT thick films prepared by the introduced method [18, 19] allow for dense piezoceramic layers with thickness between 20 μm–150 μm and line width from 10 μm to the mm–range on various substrates like Silicon and fired ZrO_2, Al_2O_3, and LTCC (Du Pont 951). Fig. 6 shows a typical SEM micrograph of a PZT thick film on an Al2O3 substrate. Fig. 7 shows a honeycomb structured array of PZT films on metalized Si-wafer.

The reaction between SiO_2 and PbO is relevant for PZT films on Si containing substrates like 96% Alumina, Si wafer and LTCC. The existence of silicon oxide in the substrate and in commercial electrode materials causes the diffusion into the PZT thick film and the reaction to Pb based silicates, which deteriorate the ferroelectric behavior of the PZT thick film. Therefore a special Au electrode was developed by Fraunhofer IKTS with modified composition preventing silicate formation during firing of the PZT thick film. For ZrO_2 and Al_2O_3 substrates a commercial Au electrode (Heraeus 5789) fired at 850°C/ 30 min can bee used.

Table 2 gives a summary of performance data of PZT thick films on various substrate materials.

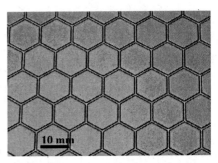

Figure 6. SEM micrograph of a polished and etched cross section of a PZT thick film

Figure 7. Honeycomb structured array of PZT thick films on metalized Si-wafer, prepared by screen printing technology.

Table 2: Properties of PZT thick films (active area a = 400 mm^2, thickness 100 µm) on various substrate materials using Au electrodes

Property	PZT on Al$_2$O$_3$ (99.7 %)	PZT on LTCC (DP 951)	PZT on Si-wafer
Dielectric constant $\varepsilon_{33}^{T}/\varepsilon_0$ at 1 kHz	1900	1500	1600
Dielectric loss tan δ at 1 kHz	0.038	0.033	0.055
Piezoelectric coefficient d$_{33}$ [pC/N]	210	180	140
Remnent polarization P$_r$ at 50 Hz [µC/cm^2]	16	10	9
Coercive field E$_C$ at 50 Hz [kV/cm]	15	13	12
Internal resistance R$_{is}$ at 30 kV/cm [Ωcm]	2x10^{11}	2x10^{11}	2x10^{11}

Piezo fiber composites - Arrange and fill process

Piezoelectric 1–3 fiber composites using the Fraunhofer methodology are prepared by epoxy infiltration of fiber bundles or fixed fiber patterns and curing and dicing of the composite. The obtained 1–3 fiber composite materials can be machined into virtually any shape or size transducer element. This "Arrange & Fill" process enables a quick, cost-efficient method of producing large quantities of composites. The piezoceramic fibers can be aligned either with or without spacer in a form and filled with polymer. Thus, regular or nonregular spacing patterns are obtainable. Especially know–how is required concerning the piezoceramic fiber fabrication, the fiber arrangement, epoxy selection, and the curing step. Originally developed at Fraunhofer IKTS, the process has been commercialized and up scaled by Smart Material Corp.[14].

Obtained piezocomposite blocks (Fig. 8) can be machined into virtually any shape or size transducer element (example, see Fig. 8). Through the use of custom molds and precision machining, concave surfaces can be produced to create a broadband sensor. The presence of polymer creates a "cushion-effect" that protects the fibers and enables the structure to resist chipping and cracking during the machining process. In guided wave inspections, where sheets or tubes type surfaces require inspection without extensive scanning, a broadband transducer is required. The piezoelectric fiber composite not only provides this desirable characteristic, but also allows for high-pulse amplitude (low noise), conformability, and an excellent acoustic impedance match to the test piece.

Figure 8. Left: piezo fiber composite block material, right: machined piezo fiber composite for transducer application

Molded composites

Fine scaled composites can be prepared starting with molded and sintered ceramic arrays made by the soft mold process. Once the piezoceramic pillars have been formed, the remaining spaces are filled with a polymer matrix material. Next the base is removed by grinding. Metal electrodes are then bonded to the ends of the fibers and first used to polarize the piezoceramic at an elevated temperature and then to apply an electric field or collect developed charges from the material. At this point, the active elements are ready to be used as piezoelectric transducer element. Experiments have shown that high-performance composites can be prepared[17].

MODULES BY MICROSYSTEMS TECHNOLOGY

Exemplary developments of microsystems as pre-integration platform in smart structure technology are described in the present paragraph. There are two promising approaches seen allowing for the integration of piezoceramic units and electronic circuits, flexible circuits boards and active ceramic multilayer devices.

Flexible circuit boards

PZT plates, preferable in the thickness range between 100 μm to 250 μm in thickness, are processed to sensor and actuator modules by packaging using flexible circuit board processing. A plenty of designs are known [20,21,22,14].

A modular concept for pre-encapsulated actuators has been developed and transferred to production by German Aerospace Center DLR Braunschweig[22]. The multifunctional elements can be designed to meet the requirements of industrial applications. This involves for example driving voltages, size and shape of the elements and the piezoceramic material itself. Recently, modules for raised operation temperatures up to 180°C have been developed. See Fig. 9. The selected materials allow for optimum actuator performance at 180 °C, see Fig. 10.

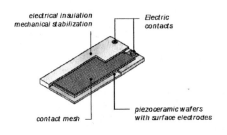

Figure 9: Sketch of the d_{31} module consisting of a monolithic layer of PZT which is embedded into epoxy, as originally developed by the German Aerospace Center[22].

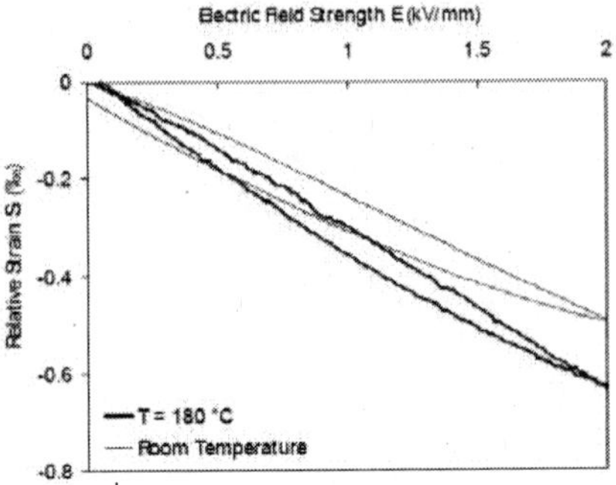

Figure 10: Measured strain characteristic of the special designed high temperature module at room temperature and 180°C

The sensor/actuator-modules modules, sketched in Fig. 9, are commercialized and are offered by PI Ceramic, Germany under the brand name DuraAct®. The transducers, typically, have an overall geometry of 40 x 20 x 0.5 mm³, the PZT plate of 30 x 15 x 0.2 mm³. Since the ceramic layer is embedded into a polymer structure, these modules offer a relative high damage tolerance; they even can be applied on curved structures.

The Macro-Fiber Composite, or MFC [21,14] is made up of rectangular uniaxially aligned ceramic fibers sandwiched between layers of adhesive and electroded polyimide films. These fibers are prepared by dicing of PZT plates using a wafer saw. Developed at NASA Langley Research Center during the late 90`s, the MFC are manufactured by Smart Material Corp. in a full-scale production, today. A variety of design forms are offered [14]. The migration from research projects to high volume, cost effective commercial applications has generated additional need for new MFC designs, electronics on microcontroller and chip level, and system design tools, as well.

By applying voltage to the MFC, the ceramic fibers change shape to expand or contract and turn the resulting force into a bending or twisting action on the material. Likewise, voltage is generated in proportion to the force applied to the MFC material. Numerous research projects have proven the concept of using the MFC in vibration and noise control applications, as well as for health monitoring, morphing of structures and energy harvesting.

Active ceramic multilayer

IKTS is now developing a piezoelectric modules made completely of ceramic[23]. The new packaging technology uses the lamination of sintered PZT ceramic plates with green low temperature co-fire ceramic layers (LTCC, Heraeus HeraLock® Tape-HL2000) and post-sintering to obtain the sensor/actuator module. This design completely integrates the PZT in the substrate material.

This approach combines LTCC microsystems technology and piezo technology and allows for a tremendous improve of functional integration, e.g. sensing, actuation, buried electronic circuits, and strain- stress transformation. The challenge exists in avoiding tension cracks at shrinking of

LTCC layers on the already sintered piezoceramic during the firing process. The advantages of a module with fully integrated PZT ceramic unit are the mechanical stabilisation of the piezoceramic, the electrical insulation and the shielding of external environmental influences. Special designed modules were integrated into Al-components by die casting. The piezoelectric modules survived this manufacturing step with full functionality, confirming the idea of adaptive metal structures in automotive and machine building industry.

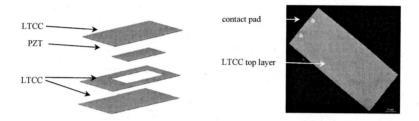

Figure 11: LTCC packaged piezo module[23]. Left: exploded view, right: prepared module

After preparation, the ceramic modules were introduced in the manufacturing chain of Aluminium die casting.

ADAPTIVE STRUCTURES
At present, a big number of research organizations (NASA, Fraunhofer, German Aerospace Center DLR, Universities) and customers (estimation a few hundred) are dealing with the use of active packaged devices for sensing, actuating, energy harvesting, health monitoring and structural control. The distribution of MFC by market segment was reported to amount to 20% defense (USA, Europe), 20% research labs, universities, 20% automotive, 15% machining equipment, 15% aerospace and 10% sporting goods, white ware, and buildings[24]. Examples of new approaches based on microsystems technologies are given in the next paragraph.

Active optical devices
The use of deformable mirrors to compensate the aberration of astronomical images caused by the turbulence of the earth atmosphere has led to outstanding successes of ground-based astronomy. In collaboration with the Active Structures Laboratory of Brussels University a demonstrator for a bimorph mirror was developed and manufactured as shown in Fig. 12 [25]. It consists of a silicon wafer with a diameter of 150 mm and 0.75 mm thickness as well as 91 honeycomb thick film piezoelectric actuators with 80 μm thickness, which have been screen printed at the backside of the silicon wafer. Each actuator can be driven individually to control the shape of the bimorph mirror and thus optical aberrations.

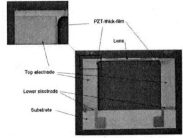

Figure 12. PZT thick film actuators with wiring for controlled driving and image correction

Figure 13. Active optical system for adjusting a FAC lens in with two degrees of freedom.

The combination of cantilever structures based on PZT thick films with solid hinges and platforms for optical devices enables for active systems with two or more degrees of freedom in deformation. In collaboration with the Fraunhofer IOF Jena, we developed a movable platform carrying the FAC (fast axis collimator) lens [26, 27]. See Fig. 13. Simultaneous driving of the two cantilevers with E = 2 kV/mm resulted in an up and down movement of the platform. Static measurements of the deflection by laser triangulation method resulted in a deflection of $\Delta l = 6$ μm. Tilting and twisting movements were possible by driving only one cantilevers.

Further investigations will build up on these experiments to open up new fields of applications for adaptive optical systems.

Active light metal components

Lightweight construction is a trend in car industry to save weight and hence to reduce fuel consumption. The use of light metals like aluminium or magnesium is one option. As general in light – weight structures, noise and vibrations becomes a problem. In 2004 we introduced the idea of direct integration of metalized and piezoelectric elements with insulating coating in a metal matrix by casting using a metallurgical fusing technique[28]. Bräutigam et al [29] investigated high pressure die casting for the preparation of active light metal structures. This technology is well established for mass production of light metal parts.

Due to the dynamic die filling and the high melt temperatures die casting is both mechanically and thermally a very challenging process for the integration of sensitive sensor/actuator-modules. For example, the velocity of the liquid metal is locally far above 100 m/s. Die filling usually takes place in less than 50 ms. As soon as the die is filled, a dwell pressure of several 100 bar is applied to compensate shrinkage and to minimize porosity. Aluminium alloys have a casting temperature within the range of 600°C to 750°C. The typical die temperature is between 150°C and 250°C.

Nevertheless, the piezoelectric LTCC-PZT modules as introduced in the preceding paragraph survived this manufacturing step without deterioration and fortify the concept of adaptive metal structures in automotive and machine building industry. Figs. 14 and 15 give an impression of the integrated ceramic module in an Aluminium plate. Die casting was done at University of Erlangen.

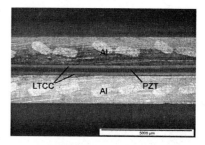

Figure 14: X-ray image (UA = 130 kV, I = 100 µA) of an aluminium die casting block with integrated LTCC piezo module[23]

Figure 15: cross-section of an integrated LTCC piezo-module with PZT- unit[23]

Development of self-powered sensor nodes

Safety of materials and structures, especially in aircraft engineering, is currently guaranteed by redundant dimensioning, periodic inspections and preventive replacement of safety relevant assemblies. Recent success in sensor - and micro electronics technology opens up the perspective of in flight inspection. Key elements are self-organizing, wireless sensor nodes with implementation of energy harvesting capability. In 2009 a mayor industrial project started in Germany, aiming at energy efficient solutions in the three focus areas computing, broadband wireless, and sensor networks[30]. One lead project deals with the development of wireless, self-powered sensor nodes using guided acoustic waves for health monitoring and life time estimation of light-weight structures. A great impact on piezo and smart structure technology can be expected in the near future from that funding.

CONCLUSIONS

Smart structure technology is going to be used for advanced products in a multitude of industrial branches. We take microsystems technologies as most promising approach for pre-integration of sensor, actuator, generator and control functions into one module. This guarantees to meet the capability of mass production at competitive cost limits. The material and design basis of these modules must be adjusted to the type of load carrying structure and the conditions of its serial production technology. As shown by example, ceramic, Si and polymer based micro system approaches are attractive starting points. The combination of multifunctional piezoceramics with electronic circuits and control and communication software in one device give tremendous potential for structures of high functionality. Let us take the continuous built-in structural health monitoring approach, rather than a traditional NDT-type maintenance approach, by self-powered sensor nodes, as example to see the possible progress of the near future.

ACKNOWLEDGEMENT

The research was funded by Germans Research Foundation DFG, SFB/Transregio 39, BMBF FKZ 13N10260 (Cool Silicon) and Fraunhofer (internal funding). The author wishes to express his gratitude to S. Gebhardt, L. Seffner, M. Flössel, U. Scheithauer, A. Michaelis, U. Partsch, V. Bräutigam, T. Daue, and R. Schmitt for essential support and discussions.

REFERENCES

[1] "Piezoelectric and Acoustic Materials for Transducer Applications", eds. A. Safari, E. K. Akdogan, Springer, 2008.

[2] Y. Saito, H. Takao, T. Tani, T. Nonoyama, K. Takatori, T. Homma, T. Nagaya and M. Nakamura, "Lead free piezoceramics", Nature, 432, 84-87 (2004).

[3] T. Tani, "Crystalline-oriented piezoelectric bulk ceramics with a perovskite-type structure", J. of the Korean Phys. Soc., 32, 1217-1220 (1998).

[4] T. Tani and T. Kimura, "Reactive-templated grain growth processing for lead free piezoelectric ceramics", Adv. Appl. Cer., 105, 55-63 (2006).

[5] T. Takenaka, K. Maruyama and K. Sakata, "(Bi1/2Na1/2)TiO3-BaTiO3 system for lead-free piezoelectric ceramics", Jpn. J. Appl. Phys., 30 [93] Part 1, No. 9B, 2236-2239 (1991).

[6] J. H. Cho, Y. H. Lee, B. J. Chu, B. I. Kim, H. T. Chung, "Processing and Properties of Piezoelectrics Based on BNT (Bi,Na)TiO3", Euro Ceramics VIII, Published in: KEM - Key Engineering Materials, 264-268, 1321-1324 (2004).

[7] S. Zhang, T. R. Shrout, H. Nagata, Y. Hiruma and T. Takenaka, "Piezoelectric properties in (K0.5Bi0.5)TiO3-(Na0.5Bi0.5)TiO3-BaTiO3 lead-free ceramics," IEEE Trans Ultrason Ferroelectr Freq Control, 54 [5], 910-917 (2007).

[8] EU-Directive 2002/96/EC: "Waste electrical and electronic equipment (WEEE)" and "RoHS, Directive on Restrictions of the Use of Certain Hazardous Substances in Electrical and Electronic Equipment" Official Journal of the European Union, 46[L37], 24-38 (2003).

[9] N.W. Hagood NW and A.A. Bent "Development of Piezoelectric Fiber Composites for Structural Actuation" Proc 43th AIAA/ASME, pp. 93–1717 (1993)

[10] K.A. Klicker et al. "Composites of PZT and Epoxy for Hydrostatic Transducer Applications" J Am Soc 64:5–9 (1981).

[11] R.B. Cass "Fabrication of Continuous Ceramic fiber by the Viscous Suspension Spinning Process" Am Ceram Bull 70:424–429 (1991).

[12] J.D. French et al., Proc SPIE – Int Soc Opt Eng 3044:406–412 (1997).

[13] E. Teager et al. "Lyocell products with build-in functional properties" Chem Fibers Int 48:32–35 (1998).

[14] www.smart-material.com

[15] U. Scheithauer Scheithauer, L. Löber, L. Seffner, A. Schönecker, M. Flössel, S. Uhlig, S. Gebhardt, A. Michaelis"Piezoceramic fibre technology based on the polysulphone process", Electroceramics XI: 11th International Conference on Electroceramics, University of Manchester, September 1-3, 2008, to be published in J Europ Ceram Soc (2009)

[16] Starke S et al. "Fine Scale Piezoelectric 1–3 Composites: A New Approach of Cost Effective Fabrication" Proc 11th IEEE International Symposium on the Applications of Ferroelectrics, ISAF'98 Vol. I: 393–396 (1998).

[17] S. Gebhardt „Herstellung und Charakterisierung von feinskaligen 1–3 Piezo-kompositen für Ultraschallwandler", Thesis TU Bergakademie Freiberg (2000)

[18] S.E. Gebhardt, T. Rödig, U. Partsch, A.J. Schönecker "Development of Micro-Intetrated Sensors and Actuators Based on PZT Thick Films", Proc. 16th Europ. Microelectronics and Packaging Conference & Exhibition EMPC 2007, Oulu, Finnland, June 17-20, pp. 177-181(2007).

[19] S. Gebhardt, L. Seffner, F. Schlenkrich, A. Schönecker, "PZT Thick Films for Sensor and Actuator Applications", Journ. Europ. Ceram. Soc., vol. 27, pp. 4177-4180 (2007).

[20]Lazarus et al., "Packaged strain actuator", U.S. Patent 5,656,882, August 1997

[21]Wilkie W.K., Bryant G.R., High J.W. et al., Proceedings of the 2000 SPIE Conference, New Port Beach, USA, March 5-9, (2000).

[22]P.Wierach, D. Sachau, A. Schönecker; ASME International Mechanical Engineering Congress and Exposition November 11–16, 2001, New York, NY, Proceedings (2001)

[23]M. Flössel, S. Gebhardt, A. Schönecker, A. Michaelis "LTCC/PZT Sensor-Actuator-Module", Electroceramics XI: 11th International Conference on Electroceramics, University of Manchester, September 1-3, 2008, to be published in *J Europ Ceram Soc* (2009)

[24]A. Schönecker, T. Daue, B. Brückner, C. Freytag, L. Hähne, T. Rödig, "Overview on Macro Fiber Composite Applications" , Proceedings of the 2006 SPIE Conference, San Diego, USA, March, (2006).

[25] http://www.ulb.ac.be/scmero/optic_segmented.html; A. Preumont, B. De Marneffe, G. Rodrigues, H. Nasser, A. Deremaeker, "Dynamics and control in precision mechanics", to be published in Revue Européenne de Mécanique Numérique.

[26]S. Gebhardt, A. Schönecker, C. Bruchmann, E. Beckert, G. Rodrigues, R. Bastaits, A. Preumont, "Active Optical Structures by Use of PZT Thick Films", Proceedings IMAPS Deutschland Technologies Conference (*CICMT*) München, 21. - 24. April, (2008).

[27]C. Bruchmann, B. Höfer, P. Schreiber, R. Eberhardt, W. Buss, T. Peschel, S. Gebhardt, A. Tünnermann, E. Beckert, "Application of PZT thick-films on adjusting purposes in micro-optical systems", Proc. SPIE Conference Microtechnologies for the New Millenium 2007, Gran Canaria, Spain, vol. 6589, pp. 65891M, May 2-4, (2007).

[28]A. Schönecker, S. Brüdgam, K. Schmidt, A. Naake, patent application, DE10315425A1

[29]V. Bräutigam et al, "Smart structural components by integration of sensor/actuator-modules in die castings" Industrial and Commercial Applications of Smart Structures Technologies, Proc. of SPIE Vol. 6527, 65270S, (2007)

[30]http://www.cool-silicon.org/index.html

PIEZOELECTRICITY IN (K,Na)NbO$_3$ BASED CERAMICS

Cheol-Woo Ahn[1], Chee-Sung Park[1], Sahn Nahm[2], and Shashank Priya[1*]

[1]Center for Energy Harvesting Materials and Systems, Materials Science and Engineering, Virginia Tech, Blacksburg, VA 24061, U.S.A.

[2]Department of Materials Science and Engineering, Korea University, 1-5 Ka, Anam-dong, Sungbuk-ku, Seoul 136-713, Korea

ABSTRACT

In this manuscript, we report the variation of piezoelectric constant (d_{33}) at room temperature in $(K_{0.5}Na_{0.5})NbO_3$ (KNN) based ceramics. Phase fraction and orthorhombic to tetragonal phase transition temperature (T_{O-T}) were found to be the controlling variables of piezoelectricity. Using the model system of $(1-x)(K_{0.48}Na_{0.48}Li_{0.04})NbO_3$-$xBaTiO_3$ [$(1-x)$KNLN-xBT] and $0.99(K_{0.48}Na_{0.48}Li_{0.04})(Nb_{1-y}Sb_y)O_3$-$0.01BaTiO_3$ (KNLN$_{1-y}$S$_y$-BT), it is shown that maximum d_{33} magnitude for KNN based polycrystalline ceramics is approximately 300 pC/N at room temperature.

Recently, lead-free piezoelectric ceramics have been widely studied as alternatives to $Pb(Zr,Ti)O_3$ (PZT).[1-14] Among the various candidates, $(K,Na)NbO_3$ (KNN) based materials have received considerable attention owing to their excellent piezoelectric properties and high Curie temperature (T_C).[1,2-12] In general, the KNN-based materials which have received the most interest for piezoelectric applications can be divided into these two categories: (i) Li-modified compositions such as KNN-LiNbO$_3$ (KNN-LN), KNN-LiSbO$_3$ (KNN-LS), KNN-LiTaO$_3$ (KNN-LT), $(K,Na,Li)(Nb,Ta,Sb)O_3$ (KNLNTS), etc., and (ii) solid solution with Ti-based ceramics such as KNN-BaTiO$_3$ (KNN-BT), KNN-CaTiO$_3$ (KNN-CT), KNN-SrTiO$_3$ (KNN-ST) etc.[1-12] Recently, we have shown that the enhancement in piezoelectric properties at room temperature in both these categories is correlated with the polymorphic phase transition between orthorhombic (O) and tetragonal (T) phase.[13-15] In this manuscript, our objective is to predict the range of piezoelectric coefficient at room temperature that can be obtained in KNN based polycrystalline ceramics by following the conventional synthesis approaches. In order to verify our hypothesis, we synthesized compositions in the systems KNN–LN–BT(KNLN-BT) and KNN–LN–LS–BT (KNLNS-BT) which combine both the categories listed above. The synthesized ceramics had similar microstructure and thus the piezoelectricity can be interpreted in terms of phase transitions.

Figure 1 summarizes the variation of longitudinal piezoelectric constant (d_{33}) at room temperature (RT) as a function of T_{O-T} in KNN based ceramics with various substituents. The data in this figure scales well with linear relationship as shown with dot line in Fig. 1. This indicates that higher piezoelectric properties for KNN ceramics are obtained by shifting T_{O-T} towards room temperature and the maximum d_{33} is in the range of 300 pC/N. Two other interesting points can be noted in this figure: (i) KNLNS-BT has lower T_{O-T} than KNLN-BT although both the composition plotted in this figure have same KNN fraction ~0.95. This is related to the fact that KNN-LS has lower T_{O-T} than KNN-LN[14]; (ii) KNLN-BT deviates slightly from the linear relationship while KNLNS-BT follows it. This implies that the magnitude of piezoelectric constant is related to additional factors.

The ceramic compositions of $(1-x)(K_{0.48}Na_{0.48}Li_{0.04})NbO_3$-$xBaTiO_3$ [$(1-x)$KNLN-xBT] and $0.99(K_{0.48}Na_{0.48}Li_{0.04})(Nb_{1-y}Sb_y)O_3$-$0.01BaTiO_3$ (KNLN$_{1-y}$S$_y$-BT), were used as model systems to illustrate the variation of piezoelectric constant. The ceramics were synthesized by conventional solid-state route using oxide powders of >99% purity. Powders of K_2CO_3, Na_2CO_3, Nb_2O_5, Li_2CO_3, $BaCO_3$, TiO_2, and Sb_2O_3 (all obtained from Alfa Aesar) were mixed for 24 h in a polypropylene jar with zirconia balls. Mixed powders were dried and then calcined at 950 °C for 3 h. Calcined powders were milled for 48 h, dried and pressed into disks under pressure of 100 kgf/cm^2 and sintered at 1080 °C for 2 h. The samples of the synthesized composition were poled in silicone oil at 120°C by applying a DC field of 4 kV/mm for 60 min. All electrical measurements were done on aged samples (24 h after poling). Figure 2 shows the X-ray diffraction (XRD) patterns of (1-x)KNLN-xBT and $KNLN_{1-y}S_y$-BT ceramics sintered at 1080 °C for 2 h (Philips Xpert Pro). All the peaks were indexed according to perovskite structure. The inset of Fig. 2 shows the variation of tetragonality in (1-x)KNLN-xBT and $KNLN_{1-y}S_y$-BT ceramics. The coexistence of O and T phase was observed for ceramics sintered at 1080 °C for all compositions. In order to observe phase variations with x and y values, the ratios of T peak intensities were calculated by the sum of the peaks intensities marked in the inset of Fig. 2.

Figure 3(a) illustrates the variation of d$_{33}$ with substitution fraction in KNN based ceramics. The fractional ratio of tetragonal peak intensity with respect to the orthorhombic one was calculated from the XRD patterns, while the tetragonality of KNN-BT ceramics was measured by Rietveld analysis as reported in our previous work.[7] We used the equation given as: $F_T=SI_T/(SI_T+SI_O)$; in order to calculate the approximate fraction of T phase (F_T), here SI_T and SI_O are the sums of peaks intensities for T and O phases. It was found that the calculated values of fractional ratio using XRD peak intensity showed a similar trend as that determined from Rietveld analysis. Thus, an approximation of the amount of the two phases (T and O) can be made by using the XRD patterns shown in Figs. 2, 3(a) and (b). T peak ratio was calculated to be 9.6% for pure KNN ceramics and 88.5% for 0.9KNN-0.1BT ceramics. Three interesting observations can be immediately made from Fig. 3(a) for KNN–BT ceramics, which are: (i) d$_{33}$ shows a maximum in a narrow range of compositions; (ii) high d$_{33}$ corresponds to T peaks ratio of 65-75%; and (iii) the maximum in d$_{33}$ corresponds to a fractional peaks ratio of approximately 70%. In conjunction with the results in Fig. 1, we can conjecture that KNN-BT ceramics will have a maximum d$_{33}$ coefficient at room temperature with T phase ratio of ~70%.

The maximum d$_{33}$ was observed for all the compositions with 5~6% substitution in KNN, as seen in Fig. 3(a) from the magnitude of KNN ratio. However, KNLNS-BT exhibits higher piezoelectric constant than KNLN-BT even though KNN ratio is around 0.95 in both cases. This enhancement, marked "I1" in Fig. 3(a), may be correlated with the lower T_{O-T} of KNLNS-BT. Even though "I1" can be explained by the variation of T_{O-T}, the increase of tetragonality (marked "I2" in Fig. 3(a)) cannot be explained using the variation of T_{O-T}. "I2" can be explained by noting the difference in ionic radii of niobium and antimony (Nb^{5+}=64 pm, Sb^{5+}=60 pm, and Sb^{3+}=76 pm).[16] The Sb-substitution increases the tetragonality of specimens and consequently the composition of KNLNS-BT ceramics was found to be located in the T-rich range.

Figure 3(b) compares the phase fractions of various compositions. The maximum d$_{33}$ was observed in the T-rich region of KNN based ceramics. In case of KNLN-BT ceramics most of composition investigated were located out of the T-rich region highlighted in the figure. For KNLNS-BT ceramics, the compositions corresponding to y=0.01 and 0.02 were located in the T-rich region. This result can explain for the higher d$_{33}$ of KNLNS-BT composition (S2) as compared to those of KNLN-BT. The maximum piezoelectric coefficient point for KNLNS-BT was found to be located in the vicinity of ~70% tetragonality. This result comes from the fact that the calculated fraction for 100% T phase is higher in KNN–LN than KNN–BT (96.8% and 88.5%, marked with S1) as shown in Fig. 3(b), and LN content is much larger (around 4 times) than BT contents in KNLN-BT and KNLNSBT ceramics. Summarizing these results, the piezoelectric properties depend on the T phase fraction and relative position of T_{O-T} with respect to room temperature. By shifting the T_{O-T} towards room temperature and adjusting the T phase fraction near 70%, the maximum in d$_{33}$ can be obtained.

Figure 4 shows the dielectric and piezoelectric properties of (1-x)KNLN-xBT and KNLN1-

ySy-BT specimens. The composition of 0.99KNLN-0.01BT showed the highest values for d33 and kp (planar coupling factor) of approximately 206 pC/N and 0.39 respectively in (1-x)KNLN-xBT ceramics, as shown in Fig. 4(a). This composition corresponds to KNN ratio of ~0.95. This result further indicates that the optimized substitution rate in KNN based ceramics is located in the range of 0.05~0.06, as shown in Figs. 3(a) and Fig. 4(a). In Fig. 4(b), where the KNN ratio of all of the compositions is around 0.95, we can clarify the effect of phase fraction on piezoelectric properties. The highest piezoelectric properties were observed when the T phase fraction was around 70%, as observed in Figs. 3(b) and 4(b). We also studied the microstructures of our samples by scanning electron microscopy. We found that all the compositions had cubical grains of similar size varying in the range of 1 to 10 μm as shown in Fig. 5. No significant change in microstructure with change in compositions was observed, excluding the possibility that microstructural effects might complicate those of phase coexistence in our observations.

In conclusion, our results show that the d_{33} variation depends on the position of T_{O-T} and T phase fraction. Using the results of Fig. 1 and Fig. 3(b), it can be predicted that the maximum d_{33} in KNN based ceramics is around 300 pC/N at room temperature (i.e. T_{O-T} is room temperature and T phase fraction is 70%).

ACKNOWLEDGEMENTS

The authors gratefully acknowledge that this study was supported by a grant from Office of Basic Energy Sciences, Department of Energy and the Fundamental R&D Program for Core Technology of Materials funded by the Ministry of Knowledge Economy, Republic of Korea. The authors would also like to thank NCFL, VT for their help in characterization.

REFERENCES
1) Y. Saito, H. Takao, T. Tani, T. Nonoyama, K. Takatori, T. Homma, T. Nagaya, and M. Nakamura: Nature **432** (2004) 84.
2) S. H. Park, C. W. Ahn, S. Nahm, and J. S. Song: Jpn. J. Appl. Phys. **43** (2004) L1072.
3) H. Y. Park, C. W. Ahn, H. C. Song, J. H. Lee, S. Nahm, K. Uchino, H. G. Lee, and H. J. Lee: Appl. Phys. Lett. **89** (2006) 062906.
4) H. C. Song, K. H. Cho, H. Y Park, C. W. Ahn, S. Nahm, K. Uchino, and H. G. Lee: J. Am. Ceram. Soc. **90** (2007) 1812.
5) K. H. Cho, H. Y. Park, C. W. Ahn, S. Nahm, H. G. Lee, and H. J. Lee: J. Am. Ceram. Soc. **90** (2007) 1946.
6) C. W. Ahn, H. Y. Park, S. Nahm, K. Uchino, H. G. Lee, and H. J. Lee: Sens. Actuators A **136** (2007) 255.
7) C. W. Ahn, H. C. Song, S. Nahm, S. H. Park, K. Uchino, S. Priya, H. G. Lee, and N. K. Kang: Jpn. J. Appl. Phys. **44** (2005) L1361.
8) Y. Guo, K. Kakimoto, and H. Ohsato: Jpn. J. Appl. Phys. **43** (2004) 6662.
9) Y. Guo, K. Kakimoto, and H. Ohsato: Solid State Commun. **129** (2004) 279.
10) Y. Guo, K. Kakimoto, and H. Ohsato: Appl. Phys. Lett. **85** (2004) 4121.
11) B. Q. Ming, J. F. Wang, P. Qi, and G. Z. Zang: J. Appl. Phys. **101** (2007) 054103.
12) H. Y. Park, K. H. Cho, D. S. Paik, S. Nahm, H. G. Lee, and D. H. Kim: J. Appl. Phys. **102** (2007) 124101.
13) C. W. Ahn, S. Nahm, M. Karmarkar, D. Viehland, D. H. Kang, K. S. Bae, and S. Priya: Mater. Lett. **62** (2008) 3594.
14) G. Z. Zang, J. F. Wang, H. C. Chen, W. B. Su, C. M. Wang, P. Qi, B. Q. Ming, J. Du, L. M. Zheng, S. J. Zhang, T. R. Shrout: Appl. Phys. Lett. **88** (2006) 212908.
15) T. R. Shrout, and S. J. Zhang: J. Electroceram. **19** (2007) 111.

16) Shannon RD: Acta Cryst. **32** (1976) 751.

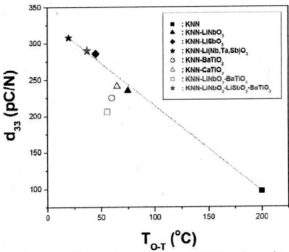

Figure 1. Piezoelectric constants (d_{33}) at room temperature in KNN based ceramics as a function of orthorhombic to tetragonal phase transition (T_{O-T}): (K,Na)NbO₃ [KNN], 0.94($K_{0.48}$Na$_{0.535}$)NbO₃-0.06LiNbO₃ [KNN-LiNbO₃], 0.948($K_{0.5}$Na$_{0.5}$)NbO₃-0.052LiSb O₃ [KNN-LiSbO₃], ($K_{0.44}$Na$_{0.52}$Li$_{0.04}$)(Nb$_{0.86}$Ta$_{0.10}$Sb$_{0.04}$)O₃ [KNN-Li(Nb,Ta,Sb)O₃], ($K_{0.5}$Na$_{0.5}$)NbO₃-BaTiO₃ [KNN-BaTiO₃], ($K_{0.5}$Na$_{0.5}$)NbO₃-CaTiO₃ [KNN-CaTiO₃], 0.99 ($K_{0.48}$Na$_{0.48}$Li$_{0.04}$)NbO₃-0.01BaTiO₃ [KNN-LiNbO₃-BaTiO₃], 0.99($K_{0.48}$Na$_{0.48}$Li$_{0.04}$) (Nb$_{0.98}$Sb$_{0.02}$)O₃-0.01BaTiO₃ [KNN-LiNbO₃-LiSbO₃-BaTiO₃]. [2,4,6-15]

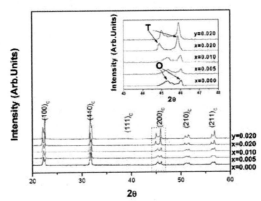

Figure 2. XRD patterns of (1-x)KNLN-xBT and KNLN$_{1-y}$S$_y$-BT specimens sintered at 1080 °C for 2 h.

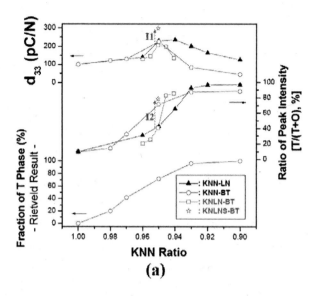

(a)

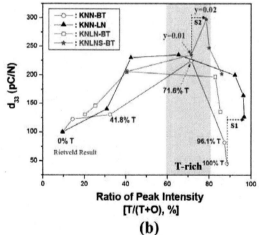

(b)

Figure 3. (a) Variations of d$_{33}$ with substitution rates, the fraction of tetragonal peaks intensities in XRD patterns, and the tetragonality measured by Rietveld analysis in KNN based ceramics, and (b) variation of d$_{33}$ with fraction of tetragonal peak intensities in XRD patterns.[7,8]

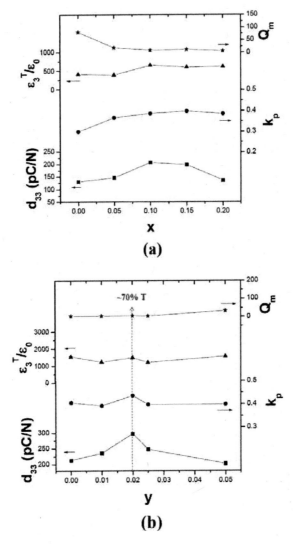

Figure 4. Piezoelectric and dielectric properties of (a) (1-x)KNLN-xBT and (b) KNLN₁₋ᵧSᵧ-BT specimens sintered at 1080 °C for 2 h.

Figure 5: All the compositions had cubical grains of similar size varying in the range of 1 to 10 μm

PHASE RELATIONSHIP AND MICROWAVE DIELECTRIC PROPERTIES IN THE VICINIY OF Ba(Zn$_{1/3}$Ta$_{2/3}$)O$_3$

Hitoshi Ohsato[1], Eiichi Koga[2], Isao Kgomiya[1] and Ken-ichi Kakimoto[1]

[1]Materials Science and Engineering, Nagoya Institute of Technology, Gokiso-cho, Showa-ku, Nagoya 466-8555, Japan. [2]Ceramic Division, Panasonic Electronic Devices Japan Co. Ltd.,1037-2, Kamiosatsu, Chitose 066-8502, Japan.

ABSTRACT

One of the complex perovskites, Ba(Zn$_{1/3}$Ta$_{2/3}$)O$_3$ (BZT) with high quality factor Q (1 / tanδ) has been used in practical application as dielectric resonator in mobile communication devices. The Q factor of BZT is affected by compositional deviation, structural ordering, microstructure, crystal phases, lattice defects and so on. In this paper, we review some results for microwave dielectric properties of BZT. Ceramics in the vicinity of BZT were synthesized by solid state reaction at 1400°C for 100 hours in air. The crystal structures were analyzed by the Rietveld method and the dielectric properties were measured by Hakki and Coleman's method. There are three regions in the vicinity of BZT. The 1st region is composed by an ordered BZT single phase. The 2nd one is by an ordered BZT with a secondary phase. The third region consists of a single phase disordered BZT. The highest Q value was obtained in the 1st region and the composition is shifted to Ba and Ta rich composition. In the second region, ordered BZT is end-member on eutectic ternary system. In the third region, the BZT is a single phase of disordered structure. Each phase in the region makes solid solutions including structural defects as detected by Raman spectra. The Qf values in the latter both cases are degraded by the presence of secondary phase and defects, respectively. Kugimiya reported dense compounds near the tie-line BMT to BaTa$_{4/5}$TiO$_3$ with high Q factor. A compound with highest Q deviated from the ideal BZT composition is also explained by Kugimiya's deviated composition.

KEYWORDS: microwave dielectric materials/properties, ordering, disorder, electroceramics, complex perovskite.

INTRODEUCTION

Complex perovskite compounds[1-4] such as Ba(Mg$_{1/3}$Ta$_{2/3}$)O$_3$ (BMT), Ba(Zn$_{1/3}$Ta$_{2/3}$)O$_3$ (BZT) have high quality factor Q which is inverse of dielectric loss tanδ, for microwave dielectrics[5]. The origin of high Q, especially relationship between high Q and ordering based on order-disorder transition, has been discussed for a long time[6-10]. The feature of complex perovskite $A(B_{1/3}B'_{2/3})O_3$ exhibit the phenomenon of the ordering of B cations.

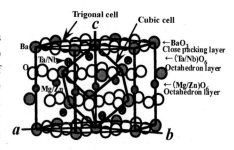

Fig. 1. Structural relationship between cubic and trigonal of complex perovskite.

Kawashima et al[11] reported that BZT has a high Q. The BZT shows ordering of B cations which is revealed by the splitting and super structure lines on the X-ray powder diffraction patterns for a long sintering time[11]. The ordering is based on the phase transition from high symmetry cubic to low symmetry trigonal. When the Zn and Ta ions occupy same position, the structure is a disordered cubic. On the other hand, if they occupy different independent sites, that is ordering, it becomes trigonal. This transition is sluggish and the temperature is not clear in some compounds. The relationship between cubic and trigonal crystal structure are shown in Fig.1. The B cations occupy the octahedra located between hexagonal closed packing layers composed of BaO$_3$. The ordering is appeared by periodic arrangement as Zn-Ta-Ta along to c-axis of trigonal. Though it is believed that the ordering brings high Q, some anti-examples appeared, that is BMT-Ba(Co$_{1/3}$Ta$_{2/3}$)O$_3$[12] and Ba(Mg$_{1/3}$Ta$_{2/3}$Sn)O$_3$[13]. Recently, Koga et al.[14-18] presented the quantification of ordering ratio using Rietveld method and the ordering state in the vicinity of BZT. Kugimiya[19] reported the composition which deviated from the BMT has high Q because of the high density composition. More recently Surendran et al.[20] showed that Ba and Mg deficient BMT compositions have high Q. In this paper, we present the phase relationship and microwave dielectric properties in the vicinity of BZT and the origin of high Q, based on Koga's data[14-18, 21].

EXPERIMENTAL

Synthesis methods of each compound as reported in previous papers are as follows: BZT samples are synthesized by Koga et al.[14,15] using solid state reaction in a container covered tightly with the lid and sintered at 1400 °C for 100 hours after decomposition of binder at 500 °C, 2 hours. Precipitated phases are identified by X-ray powder diffraction (XRPD)[15], and observed by backscattered electron image (BEI) using scanning electron microscope (SEM) equipped X-ray microanalyer (XMA)[18]. The ordering ratio was obtained by Rietveld method[22] using the structural data of ordering and disordering models[14].

Kugimiya[19] synthesized BMT compounds by master-batch method for precise control of composition (within 0.05 %). The desired compositions are mixed using four cornered compositions which are prepared before using high purity raw materials of more than 99.9 % purity. The mixtures are calcined after ball milling in alcohol. BMT ceramics with cylindrical shape are sintered at 1600 °C for 20 hours in air atmosphere. Surendran et al.[20] synthesized nonstoichiometric compositions based on Ba(Mg$_{1/3-x}$Ta$_{2/3}$)O$_3$ and Ba$_{1-x}$(Mg$_{1/3}$Ta$_{2/3}$)O$_3$ by conventional solid-state ceramic route.

The characterizations of crystalline states also presented the previous papers. Densities of these compounds were measured by Archimedes method. Microwave dielectric properties were measured using Hakki and Colman method[23, 24].

RESULTS AND DISCUSSION
Ordering ratio and Q[14]

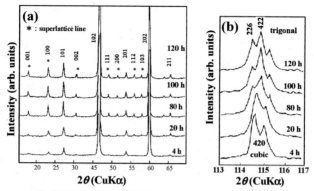

Fig.2. XRPD patterns of BZT as a function of sintering time[14].

Fig. 2 shows the XRPD patterns (a) and high angle diffraction peaks (b) of BZT ceramics as a function of sintering time at 1350 °C. According to sintering time, superlattice lines with asterisk became clear and the 420 cubic diffraction peak splits gradually into two peaks 226 and 422 in the trigonal system. It is considered that ordered and disordered structures are coexisted and ordered peaks become intense on sintering for 80 hours or more. These results are consistent with the report of Kawashima et al.[19].

The ordering ratios of BZT obtained by Rietveld method[22] are shown in Fig. 3 as a function of sintering time. The values of ordering ratio are saturated at about 80 % on sintering for 80 hours or more. Figs. 4 show Qf as functions of ordering ratio (a), density (b) and grain size (c). The ordering ratio saturates at about 80 % but the Qf varies from 40,000 to 100,000 GHZ. However, the Qf increases with density and grain size. This indicates that ordering is not so important on the Q value.

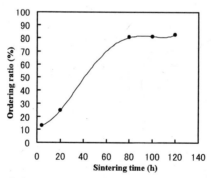

Fig. 3. Ordering ratio on BZT as a function of sintering time.[14]

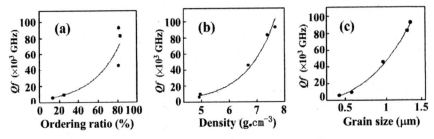

Fig. 4. Qf as a function of ordering (a), density (b) and grain size (c).[14]

Phase relation and microwave dielectric properties in the vicinity of BZT[15,18]

Koga et al.[15, 18] studied the phase relation in the vicinity of BZT in the BaO-ZnO-TaO$_{5/2}$ ternary system as shown in Fig. 5. Fig. 6 shows XRPD patterns of compositions A to S in the three series ① to ③ presented in the Fig. 5. High angle diffraction patterns around 114 to 115° on 2θ show order or disorder by peak splitting. These diffraction patterns are fitted well as shown in Fig. 7 by Rietveld method[22]. Ordering ratios obtained are shown in Fig. 8(a). Following three areas in the vicinity of BZT are presented as shown in Fig. 5 which is arranged according to Kugimiya' results which will be explained later[19].

(I)   Ordering area with BZT single phase
(II)   Ordering area with secondary phase
(III) Disordering area with BZT single phase

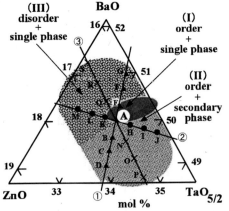

Fig.5. BaO-ZnO-TaO$_{5/2}$ ternary system in the vicinity of BZT with synthesized compositions A to S in the three series ①, ② and ③.

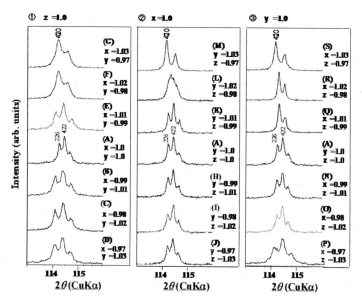

Fig. 6. XRPD patterns of composition A to S as shown in Fig. 5.[15]

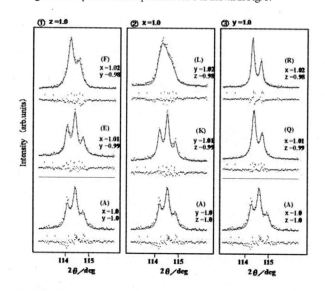

Fig. 7. XRPD patterns in the vicinity of BZT refined by Rietveld method.[15]

The first area (I) is composed of a single phase of BZT with ordered structure, and high Qf. Composition E and K have about 50 % higher Qf than pure BZT composition A. The composition K is located on the boundary area (I) and has a minor secondary phase as revealed by SEM figure as reported in a previous paper[18]. The ordering ratio in the E is lower than A, and the density of the composition E is same as that of A. The second area (II) is an ordered BZT with the secondary phase BaTa$_2$O$_6$ with certain amount of Zn[18] analyzed by XMA. The ordering ratio of compounds located in this area is high at about 70 to 80 % as shown in Fig. 8(a). Although the structure is ordered, the Qf values decrease according to deviation from pure BZT as shown in Fig. 8(b). The composition of the ordered BZT compounds is located on Ta$_2$O$_5$ rich side, which is precipitated with secondary phase as a eutectic phase diagram system. The third area (III) is precipitated as a phase single BZT solid solution with a disordered structure. The Qf values are degraded with decrease in the ordering ratio and density as shown in Fig. 8(c). The lower density comes from existents of many pores due to hard sintering. The single phase in this area is originated by solid solution accompanying defects in B- and O-sites, which introduces degradation of Qf. The pores and defects were examined by SEM[18] and Raman scattering spectra[21], respectively.

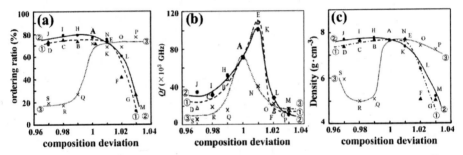

Fig. 8. Ordering ratio (a), Qf (b) and density (c) as a function of composition deviation in the vicinity of BZT.[13]

High Q by high density composition[19,20]

Kugimiya[19] presented the highest Qf composition at Ta and Ba rich side in BMT system as shown in Fig. 9. Here, chemical formulae in the vicinity of BMT are reported as follows:

Kugimiya presented three area divided by following two lines as shown in Table 1, and Fig. 9.

$$\alpha = 5\gamma/4$$
$$\alpha = \gamma/2$$

Here, α and γ are in Ba$_\alpha$Ta$_\gamma$O$_{\alpha+5\gamma/2}$.

In the region $\alpha > 5\gamma/4$, the composition denoted by Ba$_{1+\alpha}$(Mg$_{1/3}$Ta$_{2/3+\gamma}$ V$_{\alpha-\gamma}$)O$_{3+\alpha+5\gamma/2}$V$_{2\alpha+5\gamma/2}$ has B- and O-site vacancies with holes and electrons. In the $\alpha = 5\gamma/4$ line, the compositions denoted by Ba$_{1+\alpha}$(Mg$_{1/3}$Ta$_{2/3+4\alpha/5}$V$_{\alpha/5}$)O$_{3+3\alpha}$ are the ideal ones without vacancies in A and O sites. B-site vacancy is neutralized without charge. The highest Qf composition locates near the line $\alpha = 5\gamma/4$ as shown in Fig. 9. The compositions in the line are ideal for microwave dielectrics, because of no oxygen defects and high density due to substitution Ta for Mg. In the region $5\gamma/4 > \alpha > \gamma/2$, the composition denoted by Ba$_{1+\alpha}$V$_{5\gamma/6-2\alpha/3}$(Mg$_{1/3}$Ta$_{2/3+\gamma}$V$_{\alpha/3-\gamma/6}$)O$_{3+\alpha+5\gamma/2}$ has defect in A- and B-sites filled with hole and electrons. In the region at $\alpha = \gamma$, the composition denoted by Ba$_{1+\alpha}$V$_{\alpha/6}$(Mg$_{1/3}$Ta$_{2/3+\alpha}$V$_{\alpha/6}$)O$_{3+7\alpha/2}$ has same amount vacancies in both A- and B-sites filled with same holes and electrons. In the region at $\alpha = \gamma/2$, the composition denoted by Ba$_{1+\alpha}$V$_{\alpha}$(Mg$_{1/3}$Ta$_{2/3+\gamma}$)O$_{3+6\alpha}$ has vacancies only in A-site with hole and in B-site with excess electrons which introduced instability. In the region $\alpha < \gamma/2$, the composition denoted by Ba$_{1+\alpha}$V$_{\gamma-\alpha}$(Mg$_{1/3}$Ta$_{2/3+\gamma}$)O$_{3+\alpha+5\gamma/2}$V$_{\gamma/2-\alpha}$ has holes in the both A- and O-sites with electrons and excess electrons in B-site which leads to an unstable crystal structure.

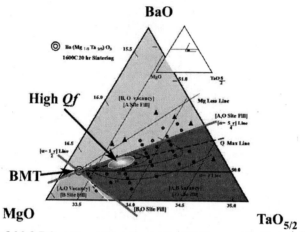

Fig. 9. BaO-MgO-TaO$_{5/2}$ ternary system in the vicinity of BMT presented by Kugimiya[19].

Table I. Chemical formulae for three areas divided by two lines: $\alpha = 5\gamma/4$ and $\alpha = \gamma/2$, here, α and γ are in Ba$_{\alpha}$Ta$_{\gamma}$O$_{\alpha+5\gamma/2}$ and V is vacancy on the A, B and O sites. (after Kugimiya[19])

α	Chemical formula	Vacancy
$\alpha > 5\gamma/4$	Ba$_{1+\alpha}$(Mg$_{1/2}$Ta$_{2/3+\gamma}$V$_{\alpha \cdot}$)O$_{3+\alpha+5\gamma/4}$ V$_{2\alpha \cdot 5\gamma/2}$	B, O: vacancy A: fill
$\alpha = 5\gamma/4$	Ba$_{1+\alpha}$(Mg$_{1/2}$Ta$_{2/3+4\alpha/5}$V$_{\alpha/5}$)O$_{3+3\alpha}$	B: vacancy A, O: fill
$5\gamma/4 > \alpha > \gamma/2$	Ba$_{1+\alpha}$V$_{5\gamma/6-2\alpha/3}$(Mg$_{1/3}$Ta$_{2/3+\gamma}$V$_{\alpha/3-\gamma/5}$) O$_{3+\alpha+5\gamma/2}$	A, B: vacancy O: fill
$\alpha = \gamma/2$	Ba$_{1+\alpha}$V$_{\alpha}$(Mg$_{1/3}$Ta$_{2/3+\gamma}$)O$_{3+6\alpha}$	A: vacancy B, O: fill
$\alpha < \gamma/2$	Ba$_{1+\alpha}$V$_{\gamma-\alpha}$(Mg$_{1/3}$Ta$_{2/3+\gamma}$)O$_{3+\alpha+5\gamma/2}$ V$_{\gamma/2-\alpha}$	A, O: vacancy B: fill

Koga's data[15] are comparable with Kugimiya's BMT data[19]. The first (I) area with highest Qf in Fig. 5 is superimposed with Kugimiya's area with high Qf as shown in Fig. 10, though the area is shifted by a small amount. The E composition in Fig. 5 will be comparable with the completed ideal crystal structure Ba$_{1+\alpha}$(Mg$_{1/3}$Ta$_{2/3+4\alpha/5}$V$_{\alpha/5}$)O$_{3+3\alpha}$ reported by Kugimiya[19]. The formula is rewritten as Ba(Mg$_{1/3-\alpha/3}$Ta$_{2/3+2\alpha/15}$V$_{\alpha/5}$)O$_3$ solid solutions on the tie-line BMT-BaTa$_{4/5}$TiO$_3$. The crystal structure on the composition region is perfect without defects and with high density. The density of BMT increases with the introduction of BaTa$_{4/5}$O$_3$ phase because Mg ions are substituted by the heavy Ta ions.

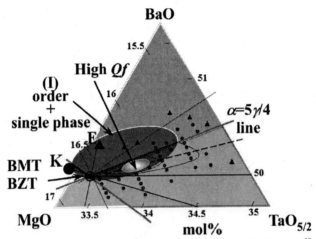

Fig. 10. BaO-MgO-TaO$_{5/2}$ ternary system in the vicinity of BMT presented by Kugimiya[19] superimposed with the first area in Fig. 5 presented by Koga et al.[15]

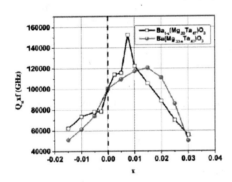

Fig. 11. Qf for Ba(Mg$_{1/3-x}$Ta$_{2/3}$)O$_3$ and Ba$_{1-x}$(Mg$_{1/3}$Ta$_{2/3}$)O$_3$ as a function of composition deviation x.[20]

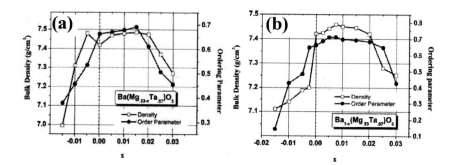

Fig. 12. Bulk density and ordering parameter of Ba(Mg$_{1/3-x}$Ta$_{2/3}$)O$_3$ and Ba$_{1-x}$(Mg$_{1/3}$Ta$_{2/3}$)O$_3$ as a function of composition deviation x[20].

Surendran et al.[20] also presented compositions with high Q on the two kinds of magnesium and barium deficient nonstoichiometric compositions in Ba(Mg$_{1/3-x}$Ta$_{2/3}$)O$_3$ [x=0.015] and Ba$_{1-x}$(Mg$_{1/3}$Ta$_{2/3}$)O$_3$ [x=0.0075] as shown in Fig.11. The microwave dielectric properties of Ba$_{0.9925}$(Mg$_{0.33}$Ta$_{0.67}$)O$_3$ [ε_r = 24.7, τ_f = 1.2 ppm/°C, Qf = 152,580 GHz] and Ba(Mg$_{0.3183}$Ta$_{0.67}$)O$_3$ [ε_r = 25.1, τ_f = 3.3 ppm/°C and Qf = 120,500 GHz] were found to be better than stoichiometric BMT [ε_r = 24.2, τ_f = 8 ppm/°C and Qf = 100,500 GHz]. The important difference from Kugimiya's results[19] is standing on the nonstoichiometry with barium or magnesium deficiency. We reconsider the Surendran's data[20] based on Kugimiya's results[19]. In the case of Mg-deficient BMT, as the composition locates near the Kugimiya's area with high Qf, the composition of the main compound must be Ba(Mg$_{1/3-a/3}$Ta$_{2/3+2a/15}$V$_{a/5}$)O$_3$ solid solutions in the tie-line BMT-BaTa$_{4/5}$TiO$_3$. As shown in Fig. 12 (a), in the solid solution area the Mg deficiency are filled with Ta and create vacancies in B-site, so that density and ordering ratio are maintained. On the other hand, the existing area of Ba-deficient BMT is included in Koga's (II) area as shown in Fig. 5, which composed with ordered BMT and secondary phase. The ordered BMT will have near composition with high density and high Qf on the BMT-BaTa$_{4/5}$TiO$_3$ tie-line presented by Kugimiya[19]. The compound by Surendran et al.[20] may be located in eutectic phase diagram region accompanying with secondary phase. But, as amount of the secondary phases is small, the detection may be difficult. Though the density and ordering ratio are maintained at a high level as shown in Fig. 12(b), Qf values steeply degrade according to the secondary phase. The compound should be stoichiometric and completed, because microwave dielectrics with high Q usually should be without defects.

CONCLUSION

In a previous study, we presented High symmetry brings High Q instead of Ordering. In this paper, we reviewed phase relationship in the vicinity of BZT for finding the reason of deviated composition on high Q, and applied that high Q comes from high dense crystal structure presented by Kugimiya[19], as follows:

Koga et al.[14] determined the ordering ratio of BZT by Rietveld method[22] and found that the ratio saturates at about 80 %. They also suggested that High Q is due to increase in density and grain size of ceramics instead of ordering.

Phase relationships in the vicinity of BZT are clarified, and following three areas are presented by Koga *et al.*[15]: (I) Ordering area with single phase, (II) Ordering area with secondary phase, (III) Disordered area with single phase. The composition with High Q is deviated from ideal BZT compound. The composition is located near the Kugimiya's reported composition[19] in the BMT-BaTa$_{4/5}$O$_3$ tie line, which is made by completely perfect crystal structure without oxygen defects. Moreover, the density is heavier, because of substitution Ta for Mg. Sebastian's group[20] also presented Ba and Mg defect composition which is also explained by Kugimiya's result[19].

For microwave ceramics, the crystal structure should be perfect with good densification in order to get good dielectric properties.

REFERENCES

[1]H. Ohsato, "New Frontiers of Microwave Dielectrics with Perovskite-Type Structure", *Bull. Ceram. Soc. Jpn*, **43**(8), 597-609 (2008) [Japanese].

[2]F. Galasso and I. Pyle, "Ordering in compounds of the A(B'$_{0.33}$Ta$_{0.67}$)O$_3$ type", *Inorg. Chem.* **2**, 482-484 (1963).

[3]S. Kawashima, M. Nishida, I. Ueda and H. Ouchi, *Tech. Rep. Inst. Electron. & Comm. Eng. Jpn.* MW80-29 (1980).

[4]S. Nomura, K. Toyama and K. Kaneta, "Ba(Mg$_{1/3}$Ta$_{2/3}$)O$_3$ Ceramics with Temperature-Stable High Dielectric Constant and Low Microwave Loss", *Jpn. J. Appl. Phsy.* **21**(10), L624-626 (1982).

[5]H. Ohsato, "Research and Development of Microwave Dielectric Ceramics for Wireless Communications", *J. Ceram. Soc. Jpn.*, **113[11],** 703-711 (2005).

[6]B. D. Silverman, "Microwave Absorption in Cubic Strontium Titanate", *Phys. Rev.* **125**, 1921-1930(1962).

[7]K. Wakino, M. Murata and H. Tamura, "Far Infrared Reflection Spectra of Ba(Zn,Ta)O$_3$-BaZrO$_3$ Dielectric Resonator Material", *J. Am. Ceram. Soc.* **69**, 34-37 (1986).

[8]T. Hiuga and K. Matsumoto, "Ordering of Ba(B$_{1/3}$B$_{2/3}$)O$_3$ Ceramics and Their Microwave Dielectric Properties", *Jpn. J. Appl. Phsy.* **S28-2**, 56-58 (1989).

[9]E. S. Kim and K. H. Yoon, *Ferroelect.* **133**, 1187 (1992).

[10]C-H. Lu and C-C. Tsai, *J. Mater. Res.* **11**(5), 1219-1227 (1996).

[11]S. Kawashima, M. Nishida, I. Ueda and H. Ouchi., "Ba(Zn$_{1/3}$Ta$_{2/3}$)O$_3$ Ceramics with Low Dielectric Loss at Microwave Frequencies", *J. Am. Ceram. Soc.*, **66**, 421-423 (1983).

[12]Y. Yokotani, T. Tsuruta, K. Okuyama and K. Kugimiya, "Low-Dielectric Loss Ceramics for Microwave Uses", *National Technical Report.*, **40**(1), 11-16 (1994) [in Japanese].

[13]H. Matsumoto, H. Tamura and K. Wakino, "Ba(Mg, Ta)O$_3$-BaSnO$_3$ High-Q Dielectric Resonator", *Jpn. J. Appl. Phys.*, **30**, 2347-2349 (1991).

[14]E. Koga, H. Moriwake, "Effects of Superlattice Ordering and Ceramic Microstructure on the Microwave Q Factor of Complex Perovskite-Type Oxide Ba(Zn$_{1/3}$Ta$_{2/3}$)O$_3$", *J. Ceram. Soc. Jpn,* **111**, 767-775 (2003) [Japanese].

[15]E. Koga, H. Moriwake, K. Kakimoto and H. Ohsato, "Influence of Composition Deviation from Stoichiometric

Ba(Zn$_{1/3}$Ta$_{2/3}$)O$_3$ on Superlattice Ordering and Microwave Quality Factor Q", *J. Ceram. Soc. Jpn.*, **113**[2], 172-178 (2005) [Japanese].

[16]E. Koga, H. Mori, K. Kakimoto and H. Ohsato, "Synthesis of Disordered Ba(Zn$_{1/3}$Ta$_{2/3}$)O$_3$ by Spark Plasma Sintering and Its Microwave Q Factor", *Jpn. J. Appl. Phys.*, **45**(9B)**,** 7484-7488 (2006).

[17]E. Koga, Y. Yamagishi, H. Moriwake, K. Kakimoto and H. Ohsato, "Order-disorder transition and its effect on Microwave quality factor Q in Ba(Zn$_{1/3}$Nb$_{2/3}$)O$_3$ system", *J. Electroceram*, **17**, 375-379 (2006).

[18]E. Koga, Y. Yamagishi, H. Moriwake, K. Kakimoto and H. Ohsato, "Large Q factor variation within dense, highly ordered Ba(Zn$_{1/3}$Ta$_{2/3}$)O$_3$ system", *J. Euro. Ceram. Soc.*, **26**, 1961-1964 (2006).

[19]K. Kugimiya, "Crystallographic study on the Q of Ba(Mg$_{1/3}$Ta$_{2/3}$)O$_3$ dielectrics", *Abstract for Kansai branch academic meeting held at Senri-Life Science,* on the *Ceramic Soc. Japan,* **B-20**, 20 (2003/9/5). *Abstract for Meeting of Microwave/Millimeterwave Dielectrics and Related Materials on the Ceramic Soc. Japan, Nagoya Institute of Technology,* Japan (2004) [Japanese].

[20]K. P. Surendran, M. T. Sebastian, P. Mohanan, R. L. Moreira and A. Dias, "Effect of Nonstoichiometry on the Structure and Microwave Dielectric Properties of Ba(Mg$_{0.33}$Ta$_{0.67}$)O$_3$ " *Chem. Mater.*, **17**, 142-151 (2005).

[21]E. Koga, H. Moriwake, K. Kakimoto and H. Ohsato, "Raman Spectroscopic Evaluation and Microwave Dielectric Property of Order/Disorder and Stoichiometric/Non-Stoichiometric Ba(Zn$_{1/3}$Ta$_{2/3}$)O$_3$", Ferroelectrics, 356,146–152, (2007).

[22]F. Izumi, and T. Ikeda, "A Rietveld-analysis program RIETAN-98 and its applications to zeolites", *Mater. Sci. Forum*, **321-324**, 198-203 (2000),

[23]B. W. Hakki and P. D. Coleman, "A Dielectric Resonator Method of Measuring Inductive Capacities in the Millimeter Range", *IRE Trans. Microwave Theory & Tech.*, **MTT-8**, 402-410 (1960).

[24]Y. Kobayasi and M. Kato, "Microwave measurement of dielectric properties of low-loss materials by the dielectric rod resonator method", *IEEE Trans. Microwave Theory & Tech.*, **MTT-33**, 586-592 (1985).

MAGNETIC PROPERTIES OF METAL-CERAMIC COMPOSITE CORE-SHELL STRUCTURES SYNTHESIZED USING COPRECIPITATION AND HETERO-COAGULATION

M. Karmarkar, R. Islam[1], C.-W. Ahn, J. T. Abiade, D. Kumar,[2] D. Viehland and S. Priya[*]

Materials Science and Engineering, Virginia Tech, Blacksburg, VA 24061.

[1]*Material Science and Engineering, University of Texas Arlington, Arlington, TX 76019.*

[2] *Department of Mechanical Engineering, North Carolina A&T State University, Greensboro, NC 27411.*

ABSTRACT

This study reports room temperature synthesis and magnetic properties of metal-ceramic composite particles. The particles constitute a core-shell structure where the core is nickel-metal, while the shell is manganese zinc ferrite (MZF). Coprecipitation was used for synthesis of MZF nanoparticles comprising the shell, whereas nickel was synthesized by hydrazine assisted reduction of nickel ions in aqueous media. A core shell structure was then obtained by hetero-coagulation to form a shell of MZF around the nickel particles. Electron microscopy and x-ray diffraction confirmed nickel cores coated by MZF shells. Magnetization studies of MZF nano-particles revealed that they were not super-paramagnetic at room temperature, as expected for such particle sizes of 20nm in size. Sintered composites of metal-ceramic particles core-shell exhibited a magnetostriction of 5ppm.

INTRODUCTION

Core-shell metal–ceramic composite particles are of significant research interest for novel properties that could be useful for various applications. They also offer the possibility of developing composite structures with multifunctional characteristics. Recently, magnetic core-shell particles have been synthesized with metal –metal oxide structures such as Co-CoO [1,2], FePt-Fe$_2$O$_3$ [3], CoFe-Fe$_3$O$_4$ [4], Ni-Fe2O3 [5,6] and FeNi$_3$-NiZnFe$_2$O$_4$ [7]; and with all metal oxide structures such as CoFe$_2$O$_4$-ZnFe2O4 [11]. These particles are considered prospective candidates for high capacity magnetic memory devices [8], microwave, EMI shielding [5,6,9] and biomedical applications [10-12]. One way to synthesize core-shell structures is by combining the process of co-precipitation with hetero-coagulation [13-16]. This would allow co-synthesis of two dissimilar material systems. The process of hetero-coagulation consists of binding the particles with opposite surface charges dispersed in a solution [15,16]. At a specific pH value, one of the species has positive surface charge density, and the other a negative one. The zeta potential is both a function of dispersing media and of pH, and can accordingly be tailored by additives. The individual particles to be hetero-coagulated can be synthesized by various means. For example ferrite particles can be synthesized by coprecipitation [17-20], sol-gel [21], spray pyrolysis [22] and hydrothermal [23] methods. Coprecipitation followed by hetero-coagulation of two dissimilar types of magnetic particles offers the possibility to develop,

---

[*]spriya@vt.edu

for example, exchange-biased core-shell composite particles that have suitable magnetic and electrical properties for spintronics.

In this study, we synthesized metal-ceramic magnetic composite material systems with a core-shell structure by using co-precipitation and hetero-coagulation techniques. We have synthesized core-shell structures that (i) have Ni particles of μm size which are near-fully coated with nano particles of MZF; (ii) have Ni particles of sub-micron size which are fully coated with nano-particles of MZF; and (iii) have resultant magnetic properties superior to those of either core or shell. The core-shell particles were subsequently sintered in a $(N_2 + H_2)$ atmosphere at high temperature to achieve dense magnetostrictive metal-ceramic composites. MZF was selected as the magnetic oxide system because of its high permeability, high resistivity and potential for electromagnetostriction (i.e., electric field controlled magnetostriction); whereas, Ni was selected as the metal system because of its ability to be co-sintered with ceramics, and its high conductivity and magnetostriction. Nickel has an anisotropic magnetostriction (λ_{111}=-25×10^{-6} and λ_{100}=-46×10^{-6}) with a <111> easy axis, and its electrical resistivity is of the order of $6 \times 10^{-6} \, \Omega \cdot$cm.

EXPERIMENTAL PROCEDURE

Ammonium hydroxide, $MnSO_4.H_2O$, $ZnSO_4.7H_2O$ and $FeSO_4.7H_2O$ (Alfa Aesar) were used as starting reagents for synthesis of MZF. Nano particles were synthesized by a coprecipitation route using individual metal salts. The initial steps consisted of hydrolysis of metal salts (sulfates, chlorides, nitrates) into their respective hydroxides; as a subsequent step, the metal hydroxide precipitates form the spinel ferrite structure on heating between 85-95°C under constant stirring: $xMn^{+2}+yZn^{+2}+(3-x-y) \, Fe^{+2}+8OH^- \rightarrow Mn_xZn_yFe_{3-x-y}O_4+ 4H_2O$. Salts of the metal sulfates ($MnSO_4.H_2O$, $ZnSO_4.7H_2O$ and $FeSO_4.7H_2O$) were dissolved in stoichiometric proportion (0.05M, 0.05M, 0.2M respectively) in deionized water and mixed with a 1M 35% aqueous NH_4OH solution preheated at 85°C under constant stirring. A higher concentration of the precipitating base was intentionally added to achieve a faster reaction time and a larger precipitate size. The stoichiometry of resulting MZF particles is sensitive to the process parameters. It has been reported that for pH values >10 and for an alkalinity >2 an stoichiometric MZF can be obtained [24-26]. Accordingly, the resulting mixture was continuously stirred, at a pH of 10 and a temperature of 85°C, for two hours. During which time the metal hydroxides reacted to form spinel ferrite. The precipitates were magnetically decanted and washed repeatedly (5 times) in deionized water until the pH of the supernatant was close to 7. The powders were dried in air at 70°C for 5 hours.

Spinel phase formation was confirmed by X-ray diffraction (Powder XRD D500, Siemens Kristalloflex 810), using Cu Kα radiation (λ= 1.54 Å) at a scan speed of 0.01°sec^{-1}, and a dwell time of 0.5 sec. The as-synthesized MZF particles were confirmed to have the spinel structure, as shown in Fig. 1(a). However, after repeatedly storing freshly synthesized particles over a period of 3 days, we found that the particles exhibited a change in chemical composition, as quantified by elemental analysis using energy dispersive analysis by X-rays or EDX (Supra 55VP Leo). The reason for this change may be related to a room temperature decomposition of the particles into iron oxide (although XRD did not reveal a new phase formation), or the leaching of ions from samples stored under wet conditions.

Nickel powders were also procured from a commercial source (Inframat Advanced Material Inc.), which were determined to have a multimodal size distribution. The commercial nickel powders were larger in size as compared to the synthesized ones and had spherical shape. Hetero-coagulation was performed by ultrasonically mixing solutions of filtered MZF precipitates and nickel particles (1% and 5% by volume, respectively) for 10 minutes in an inert atmosphere to avoid oxidation of nickel. This procedure was repeated at various pH levels over the range of 2<pH<12, which was controlled by HNO_3 (EM Science) or tetra ethyl ammonium hydroxide (TEAOH, Alfa Aesar) in deionized water. Figure 1(b) shows XRD patterns for the core-shell composite particles. In this figure, the Ni peaks over shadow the MZF ones. The crystallite sizes of the MZF shells is too small to yield sharp peaks and hence are quite broadened, whereas the Ni particles are larger and well-crystallized. Please note in this figure that the presence of MZF which coated Ni was magnified by a log intensity plot. Nickel particles synthesized by aqueous hydrazine reduction were used for hetero-coagulation with MZF using the same procedure. The core-shell structure was directly confirmed by transmission electron microscope or TEM and by selected area electron diffraction pattern or SAED (JEOL 1200EX).

Zeta potential measurement was done by using Malvern Nano ZS ZEN3500 zetasizer. The sample solution was synthesized by dispersing 1-5 mM of particles under consideration in 1mM of NaCl and DI water followed by automated stirring at 80 rpm. Concentration of particles in the solution was decided by utilizing the phase data, standard deviation of zeta potential and stability of particles against self agglomeration. The magnitude of pH was varied from basic to acidic scales by addition of 1M NaOH and 0.1M HCl solutions.

RESULTS AND DISCUSSION

The extent of hetero-coagulation was monitored at each pH by SEM/ EDX analysis and TEM/SAED. Figure 2 shows typical SEM images that illustrate the effect of pH on the extent of MZF coating over the nickel cores (nickel particles obtained from commercial source). It can be seen that a pH of 10 yields the most uniform MZF coating. EDX analysis of the particles, shown in the images of Figs. 2(a) – (d), was then performed. Studies of uncoated particles (Fig. 2a) revealed the presence of only nickel, whereas coated ones had signatures of both MZF and Ni, via their respective electron excitation energies. Figures 3(a) – (c) shows zeta potential measurements on Ni and MZF particles. The small zeta potential for commercial Ni might be explained on the basis of Fig. 3(d), which shows a wide distribution of particles sizes. The large particles may act as centers of attraction for smaller Ni particles, forming agglomerates that internally shield the solution, giving rise to small zeta potential values (ζ=q/$4\pi\varepsilon\varepsilon_0$a, where a is the radius of particle and q is the charge). Comparatively, Ni particles synthesized using hydrazine assisted reduction had a smaller size range over a narrower interval, and thus higher zeta potentials, as shown in Fig. 3(c). Large negative zeta potential values for MZF are due to the basic conditions used for MZF synthesis (pH>10), unreacted metal hydroxides and presence of broken OH bonds on the surface of the particles. Due to large differences in surface charge between MZF and commercial Ni particles in the pH range of 7 to 12, hetero-coagulation can be expected to occur in this region.

The structure of core-shell composite particles was then characterized by TEM. Figure 4 shows a TEM image for fully MZF coated Ni particles: the dark spherical regions are the nickel cores, whereas the lighter areas are the MZF shell coatings. The MZF shell particle size can be seen to be ~20 nm, and an electron diffraction pattern taken from a region containing only MZF particles (marked by an arrow in Fig. 4) is shown in the inset. This diffraction pattern revealed rings, which is consistent with the broad XRD peaks, reflecting the small size of the nano-particles. The zones of various rings are identified in the inset. The ring of the strongest intensity was indexed to be 2.56 Å, which corresponds to the (311) reflection of the MZF spinel structure. The combinations of Figs. 2 and 4 directly confirm that MZF nanoparticles can be near uniformly coated on larger Ni particles, forming a magnetic oxide / metal core-shell structure.

The magnetic properties of as-synthesized particles were then studied after drying using an alternating gradient magnetometer or AGM (Princeton Measurements Corporation, Micromag 2900). The MZF particles were found to have a coercivity of H_c=45Oe, which is higher than the H_c=1Oe of sintered bulk MZF [23-27], as illustrated in Fig. 5a. The saturation magnetization of coprecipitated MZF was M_s=54emu/gm, which is lower than the M_s=60-80 emu/gm for bulk samples (milled, calcined and sintered): possible reasons are mentioned below. The magnetization of pure nickel is given in Fig. 5(b), and is similar to previously reported values [29]. Since MZF is an iron based nanocrystallite, it has a higher normalized saturation magnetization, relative to pure Ni. Next, we show the magnetic properties of Ni-MZF core-shell structures. In Fig. 5(c), it can be seen that these composite particles had lower values of H_c than pure Ni, while having a M_S comparable to that of MZF. Furthermore, the magnetic properties of these core-shell structures were found to be dependent on the volume fraction of the Ni and MZF phases; although, we have yet to perform a detailed investigation to determine the form of any composite rule of mixing.

The lower magnetizations for MZF nano particles may be attributed to (i) non–stoichiometry: if MZF deviates from ideal stoichiometry, the magnetization will be reduced by Fe_2O_3 2nd phase precipitation; (ii) moisture: since our particles were dried at 100 °C, there may be up to 10wt% residual H_2O that does not dehydrate until heating above 250°C; and /or (iii) small particle size: decreasing particle size has been previously reported to reduce the magnetization [30]. In addition, the formation of surface dead layers [31], non-saturation effects due to a random particle size distribution [32], and changes in normal cation distribution [24-18] have all been reported to reduce the magnetization of nanoparticles. It should be noted that MZF particles are nanosized but are not superparamagnetic, since they do show coercivity [33], as will be discussed in more detail below.

As shown in Fig. 3(d), the commercial nickel particles were found to have a large particle size variation by SEM particle size analysis: 50<d< 20,000 nm. Such a wide variation of particle size can have a direct influence on hetero-coagulation with MZF. Hence, submicron size nickel particles were synthesized in the lab by reduction of nickel ions (nickel sulfate hexa hydrate-98%) with hydrazine monohydrate and NaOH in aqueous solution [34,35]. The resulting particle size distributions was more narrow and had a smaller average sizes of 300<d<500 nm, as can be seen in the SEM image of Fig. 6(a). The roughness of the nickel particle surfaces is characteristic of hydrazine and nitrogen reductions of particles formed in the solution. The SEM image in Fig 6(b) shows composite particles synthesized by hetero-coagulation with Ni. Following the steps described above, an optimum coating was found for a pH in the range of 7 – 8. The MZF coating was nearly complete, although its topology was not uniform. It can be seen in the TEM image of Fig. 6(c) that the surface roughness of particles is accentuated by hetero-coagulation. Please note that the nickel particles were too thick for electron transmission, hence the diffraction pattern reveals inter-planar spacings (rings) only for the MZF coatings.

The magnetization (M) of these composite particles was then measured as a function of temperature (T) by SQUID as shown in Fig. 7. It is known that the susceptibility of nanoparticles under zero-field cooling reaches a maximum at the blocking temperature, given as: $T_b = KV/k_B \ln(t_m/\tau_o)$, where K is the magnetocrystalline anisotropy, V is the nanoparticle volume, KV is the energy barrier for reorientation of the magnetic moments of the nanoparticles, t_m is the measuring time, τ_o is an attempt frequency ($\sim 10^{-9}$s), and k_B is the Boltzmann constant. The blocking temperature deduced from the M-T response under a field of 1000Gauss was about 40K for MZF, 70K for Ni and 100K for Ni-MZF composite particles. Following the Néel-Brown relaxation model, the blocking temperature T_B corresponds to the system's transition from a weakly ferromagnetic state ($T<T_b$) characterized by remenence and hysteresis to a superparamagnetic state ($T>T_b$) characterized by thermally-assisted magnetization fluctuations between symmetry equivalent variants. The room temperature M-H hysteresis measurements are shown in Fig. 7(d) – (e). The M-H plots for these three types of particles illustrate that the onset of superparamagnetic behavior (characterized by a slim M-H curve) occurs at a temperature close to T_b. The higher field will shift T_b to lower temperatures as we see in the M-T plot. It can also be seen in these figures that the change of the hysteresis with temperature for Ni and Ni-MZF composite particles was similar in nature. Qualitatively, this demonstrates success in achieving well coated core-shell metal-ceramic particles. The next step will be to achieve bulk composites by using these particles and quantify the magnitude of magnetostriction.

Next, differential scanning calorimetry (DSC) and thermal gravimetric analysis (TGA) measurements were done on mixtures consisting of 20% core-shell MZF-Ni particles and 80% coprecipitated MZF ones. These measurements were done in an inert atmosphere of helium with a flow rate of 20ml/min and a heating rate of 10K/min. Analysis revealed a nickel oxidation peak between 525-575°C, even though the system was under an inert atmosphere: this is because the MZF particles surrounding the nickel are an oxide phase and thus can provide oxygen for nickel oxidation, via their own reduction. We found that a reducing atmosphere was necessary for maintaining metallic nickel in sintered composites. Previously Drofenik et al. reported that the phase stability of coprecipitated MZF is sensitive to sintering conditions such as the rate of heating, rate of gas flow, and sintering temperature above 1000°C that reduces the density due to Fe_3O_3 decomposition.[22] In our studies, the composite particles were dried and pressed using a cold isostatic press (CIP) under a pressure of 193MPa, and sintered in a mild reducing environment of 5% H_2 and 95% N_2 at a flow rate of 110 ml/min in a tube furnace at a temperature of 900°C. Figures 8(a) and (b) show SEM micrographs of sintered samples, which include Ni and Fe elemental maps that indicate the Ni remains in the metallic state after sintering. We found that the sintered samples had voids and porosity in the interior, as shown in the FIB cut cross-section in Fig. 8(c). We refer to these as interior because they were not observed in polished surfaces by SEM. The reason for internal voids may be the agglomerating nature of coprecipitated nanopowders of MZF during sintering[21-22].

Finally, magnetostriction measurements were performed on sintered samples containing 20wt% MZF-Ni composite particles and 80 wt% co-precipitated MZF as shown in Fig. 9(a). The samples were discs that were 5.6mm in diameter, with a thickness of 0.60mm. A strain gauge (120ohm: EA-06-060LZ-120) was bonded to the polished surfaces of sintered samples, and the strain gauge response was measured via a Wheatstone bridge circuit as shown in Fig. 9(b). The value of magnetostriction was found to be 5.2ppm which is on the high end of that typical for $(Mn,Zn)Fe_2O_4$ bulk materials (.02-5ppm)[25,26], but lower than that of pure nickel (-40 ppm). However, our samples had high MZF particle fractions. These results demonstrate success in synthesizing metal-ceramic core-shell composite structures, which are also attractive for electromagnetostrictive structures. A dense microstructure with a more homogenous phase distribution might help in enhancing the magnetostriction.

CONCLUSION

In summary, this study reports the synthesis of Ni-MZF core&shell structures by coprecipitation followed by hetero-coagulation. Structural investigation by XRD and TEM revealed a near full coverage of Ni core particles by MZF nano-particle shells. The composite particles synthesized using commercial Ni particles, had low coercive fields similar to Ni, but higher saturation magnetizations similar to MZF suggesting a means to uniquely control magnetic and electrical properties by core-shell design. The composite particles synthesized using submicron-nickel particles had properties similar to that of nickel. A sintered composite comprising of large volume fraction of MZF particles exhibited magnetostrictive strain of 5 ppm.

Acknowledgement: This research is fully sponsored by National Science Foundation through DMR – Metals.

REFERENCES:

1) V. Skumryev, S. Stoyanov, Y. Zhang, G. Hadjipanayis, D. Givord, J. Nogués, (2003) Nature 423, 850.
2) H.-G. Boyen, G. Kästle, K. Zürn, T. Herzog, F. Weigl, P. Ziemann,O. Mayer, C. Jerome, M. Möller, P. Spatz, M. Garnier, P. Oelhafen (2003)Adv. Funct. Mater. 13, 259.
3) Hao Zeng, Jing Li, Z. L. Wang, J. P. Liu, and Shouheng Sun (2004)Nano letters, V 4 No1, 187
4) Jing Li, Hao Zeng, Shouheng Sun, J. Ping Liu, and Zhong Lin Wang (2004) J. Phys. Chem. B, 108, 14005.
5) C Brosseau and P Talbot (2005) J. Appl. Phys. 97, 104325.
6) Christian Brosseau, Stéphane Mallégol, Patrick Quéffélec and Jamal Ben Youssef (2007) J Appl. Phy. 101,034301.
7) X Lu, G Liang, W Zhang (2007) Nanotechnology 18,015701.
8) S.A. Majctich, Y. Jin (1999) Science 284,470.
9) O Yavuz, M Ram, M Aldissi, P Poddar, S Hariharan (2005) J Mater. Chem. 15, 810.
10) P Gangopadhyay, S Gallet, E Franz, A Persoons, and T Verbiest (2005) IEEE Trans. Mag. 41, 10.
11) Y Qiang, J Antony, A Sharma, J Nutting, D Sikes and D Meyer (2006)J of Nanoparticle Research 8, 489.
12) Q. Zeng, I. Baker, J. A. Loudis, and Y. Liao, P. J. Hoopes and J. B. Weaver (2007) Appl. Phy. Lett. 90, 233112.
13) K. Furusawa and C. Anzai (1987) Col. Poly. Sci. 265, 882.
14) F. Tang, H Fudouzi, T Uchikoshi, T. Awane and Y Sakka (2003) Chem. Lett. 32(3), 276.
15) F. Tang, H. Fudouzi, J. Zhang and Y. Sakka (2003) Scripta Mat. 49, 735.
16) J. Lee, S. Hong. J. Lee, Y. Lee and J. Choi (2004) J. Mater. Res. 19, 1669.
17) B. Jeyadevan, C N Chinnasamy, K Shinoda, K Tohji, and H Oka (2003)J Appl. Phy., v 93, n 10, 15, 8450

18) B. P Rao, C O Kim, C G Kim, I. Dumitru, L. Spinu, and O. F. Caltun (2006). IEEE Trans. Mag. V 42 10.
19) R. Arulmurugan, B. Jayadevan, G. Vaidyanathan and S. Sendhilnathan (2005) J Mag. and Mag. Mats. 288, 470.
20) S Yan , W Ling and E Zhou (2004) Journal of Crystal Growth,v 273, n 1-2, 17, 226.
21) X Zhao, B Zheng, H Gu, C Li, S C Zhang and P D Ownby (1999) Journal of Materials Research, v 14, n 7, 3073
22) W F Kladnig and M F Zenger (1992) J Eur. Cera. Soc., v 9, n 5, p 341.
23) A. Goldman (2006) Modern Ferrite Technology 2nd edition, NY Springer, 266.
24) E. Auzans (1999) Mn-Zn ferrite nanoparticles for water- and hydrocarbon-based ferrofluids: preparation and properties. PhD thesis, Institute of Physics, University of Latvia.
25) E Auzans, D Zins, E Blumsm and R Massart (1999) J. Mat. Sci. 34, 1253.
26) M. Rozman and M. Drofenik (1995) J. Am. Ceram. Soc. 78, 2449.
27) E. C. Snelling, Soft Ferrites-properties and applications (1969) London Iliffe books Ltd, 43.
28) M. Ma, Y. Zhang, X. Li, D. Fu, H. Zhang and N. Gu (2003) Col. Surf. A: Phys. Eng. Aspects 224, 207.
29) NIST certificate, Nickel standard reference material- 762 for magnetic moment.
30) K Mandal, S. Chakraverty, S P Mandal, P Agudo, M. Pal and D. Chakravorty (2002) J App, Phy. V 92, 1.
31) Z. X. Tang, C. M. Sorensen, K. J. Klabunde, and G. C. Hadjipanayis (1991) Phys. Rev. Lett. 67, 3602.
32) J. M. D. Coey (1971). Phys. Rev. Lett. 27, 1140.
33) C. Rath, N. Mishra, S. Anand, R. Das, K. Sahu, C. Upadhyaya, and H. Verma (2002) J Appl. Phy. 41, 2211.
34) D. Chen, C. Hsieh (2002) J. Mater. Chem., 12,2412.
35) J Park, E. Chae, S. Kim, J. Lee , J. Kim, S. Yoon and J, Choi (2006) Mat. Chem. And Phy. 97, 371.

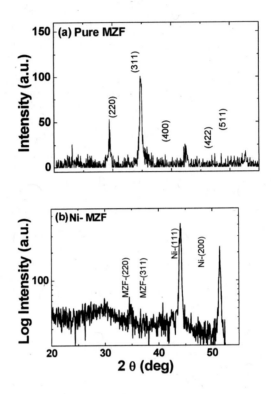

Figure 1:   X ray diffraction pattern of pure (a) MZF precipitates and (b) Ni coated with MZF

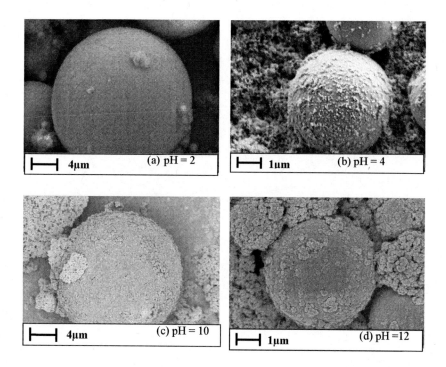

Figure 2:   SEM micrographs of extent of composite coating of MZF shell over Nickel core as a function of pH; (a)pH=2,(b)pH=4,(c)pH=10 and (d) pH=12

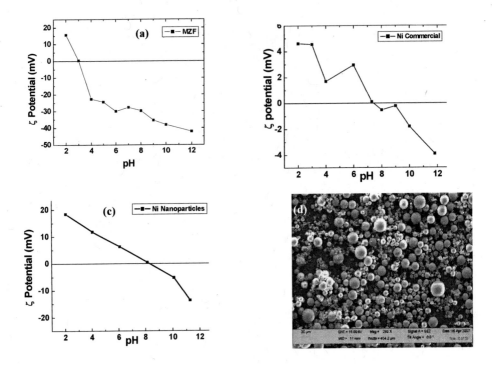

Figure 3:   Zeta potential measurement on (a) MZF nanoparticles, (b) commercial nickel, (c) nickel particles

synthesized by hydrazine reduction. (d) SEM image of the commercial nickel particles.

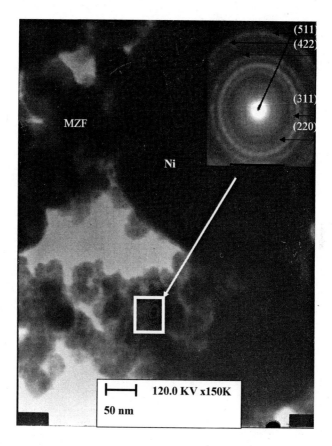

Figure 4:   TEM image of Ni coated with MZF with inset showing indexed SAED pattern of MZF and the area corresponding to the SAED.

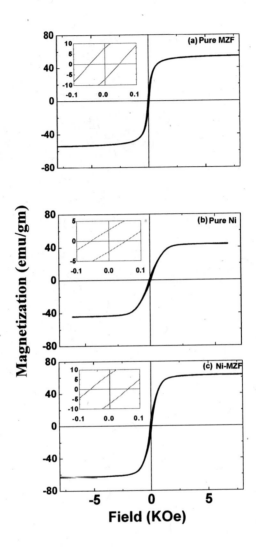

Figure 5: Magnetization measurement of (a) pure MZF precipitates (b) composite MZF coated Ni (c) Pure Ni particles. Inset in each figure shows the magnified loop at low fields to read coercivity and remenence magnetization.

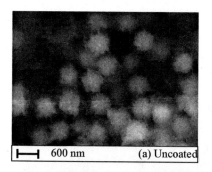

(c) TEM of particles in (b)

Figure 6:   (a) Pure nickel particles synthesized by hydrazine reduction of nickel ion, (b) Hetero-coagulated

nickel MZF particles at pH 8, (c) TEM image of MZF coated Ni with inset showing SAED pattern of MZF.

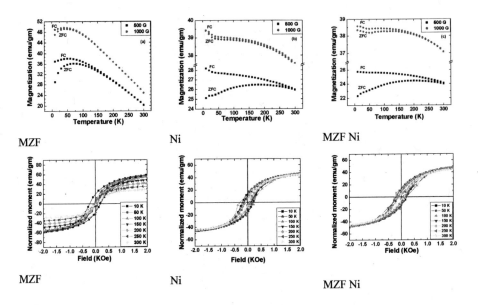

MZF   Ni   MZF Ni

MZF   Ni   MZF Ni

Figure 7:   Magnetization measurement of MZF, Ni and core-shell Ni MZF for lab synthesized nickel.

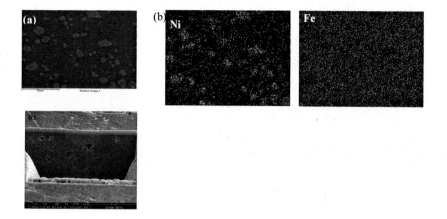

Figure 8:   (a) sintered SEM micrograph with (b) elemental mapping showing nickel and iron (denoting

MZF) regions and (c) Focused Ion Beam cut away view of sintered sample across a nickel

particle.

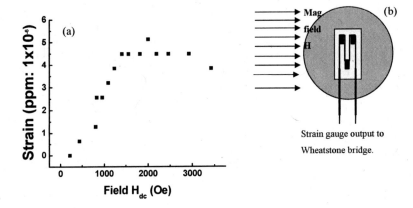

Figure 9:   (a) Strain measured as a function of magnetic field and (b) measurement setup for magnetostriction.

STRONG ELECTRIC FIELD TUNING OF MAGNETISM IN MULTIFERROIC HETEROSTRUCTURES

Ming Liu, Jing Lou, Ogheneyunume Obi, David Reed, Carl Pettiford, Stephen Stoute and Nian X. Sun

Department of Electrical and Computer Engineering, Northeastern University, Boston, MA, USA
E-mail: (nian@ece.neu.edu)

ABSTRACT

Magnetoelectric coupling can be realized in multiferroic composites, which has led to novel devices, such as the electrostatic tunable microwave signal processing devices, magnetic field sensors, etc. The performance of these devices based on multiferroic composites is critically dependent on the strength of the ME coupling, which however has been weak in at microwave frequencies. In this work, we report on novel multiferroic heterostructures which show strong ME coupling at microwave frequency. Electric field induced large changes in ferromagnetic resonance (FMR) frequencies were observed in FeGaB/Si/PMN-PT (lead magnesium niobate-lead titanate) multiferroic heterostructures, which exhibited a large tunability of the FMR frequency of Δf =900 MHz or $\Delta f/f$ = 58%. We have also investigated a whole series of spin-spray deposited ferrite/ferroelectric heterostructures, including $Ni_{0.23}Fe_{2.77}O_4$(NFO)/PZT (lead zirconium tinatate), $Ni_{0.26}Zn_{0.1}Fe_{2.63}O_4$(NZFO)/ (011) cut PMN-PT, $Zn_{0.1}Fe_{2.9}O_4$(ZFO)/ (011) cut PMN-PT, Fe_3O_4/PZT, Fe_3O_4/ (011) cut PMN-PT, and Fe_3O_4/ (011) cut PZN-PT (lead zinc niobate-lead titanate). Strong atomic bonding was observed at the interface of the spin-spray deposited ferrite/ferroelectric heterostructures, which could lead to strong ME coupling. Electric field induced giant magnetic anisotropy fields were observed in these ferrite/ferroelectric multiferroic heterostructures, which resulted in a giant electrostatically tunable FMR field range of ΔH_r=860Oe in Fe_3O_4/PZN-PT, corresponding to a large microwave ME coefficient of 108 Oe cm/kV. Static ME interaction was also investigated for these multiferroic heterostructures, which agreed well with the ME coupling at microwave frequencies. In addition, a significantly enhanced tunable ferromagnetic resonance field of 1450 Oe was demonstrated in the Fe_3O_4/ (011) cut PZN-PT heterostructure by utilizing the anisotropic piezoelectric coefficient of the (011) cut single-crystal PZN-PT. These novel multiferroic heterostructures with giant electric filed induced tunable ferromagnetic resonance fields and ferromagnetic resonance frequencies provide great opportunities for electrostatically tunable microwave multiferroic devices.

1. INTRODUCTION

Tuning of magnetization is of great fundamental and technological importance. Conventionally, magnetization tuning in many microwave magnetic devices is performed by external magnetic fields generated by electromagnets, which is slow, bulky, noisy, and energy consuming and puts severe limits on the applications of these magnetic devices. Recently, multiferroic composite materials having two or more ferroic (ferroelectric, ferro/ferrimagnetic, etc.) phases have drawn an increasing amount of interest due to their potential applications in many multifunctional devices [1-14]. Such materials can display a large stress/strain mediated magnetoelectric (ME) effect, *i.e.* a dielectric polarization variation as a response to an applied magnetic field, or an induced magnetization by an external electric field. Electrostatic control of magnetization, if realized, will lead to new magnetic devices that

53

are fast, compact and energy efficient, and can prevail in a wide variety of applications, such as information storage, sensors, RF/microwave systems, etc.

Exciting results have been reported recently on novel multiferroic composite materials and devices, such as ultra-sensitive magnetic field sensors [16], tunable resonators [17], tunable inductors [15], phase shifters [18], tunable band-stop filters [19] and tunable band-pass filters [20], etc., which provide great opportunities for mobile communication systems. However most of these tunable microwave multiferroic devices show very limited electrostatically tunable frequency range of 30 ~ 120 MHz at operation frequencies of 3 GHz or above and a limited tunable magnetic field of ~50 Oe, which severely limit their applications.[18-20] In order to improve the performance of microwave tunable devices, the magnetic material in the multiferroic composite needs to possess a large saturation magnetostriction constant, a low saturation magnetic field and a narrow ferromagnetic resonance linewidth. In this work, metallic FeGaB magnetic films with high saturation magnetostriction constant, low saturation field and low FMR linewidth were applied in FeGaB/Si/PMN-PT multiferroic heterostructures, in which a high tunable FMR frequency of $\Delta f=900MHz$ or $\Delta f/f=58\%$ was observed. In addition, spin-spray deposited ferrite/ferroelectric multiferroic heterostructures were investigated, which show giant tunable field range of 860 Oe, corresponding to ME coefficient of 108 Oe cm/kV. The large tunable FMR frequency and FMR field in these novel multiferroic heterostructures provide great potential for electrostatically tunable microwave multiferroic devices.

2 MAGNETOELECTRIC COUPLING IN FeGaB/Si/PMN-PT HETEROSTRUCTURE

Most recently, we have reported a new class of microwave magnetic thin film materials, the FeGaB films,[21] which showed excellent magnetic softness with coercivity <1 Oe, low saturation field of ~20 Oe, narrow FMR linewidth of 16~20 Oe at X-band (9.6 GHz), large saturation magnetostriction λ_s of 70 ppm, high saturation magnetization of 11~15 kG, and a self-biased FMR frequency of 1.85 GHz. The combination of these properties makes the FeGaB films excellent candidates for tunable ME microwave devices and other RF/microwave magnetic device applications. By using DC magnetron sputtering technique, amorphous $Fe_{75}Ga_{10}B_{15}$ (FeGaB) film with a thickness of 100 nm was deposited on (001) Si substrate which was then epoxy glued onto [011] cut lead magnesium niobate lead titanate (PMN-PT) to form multiferroic FeGaB/Si/PMN-PT heterostructure. (011) cut PMN-PT single crystals show anisotropic in-plane piezoelectric coefficients of d_{31} and d_{32}, with a negative d_{31} of -1800 pC/N and a positive d_{32} of 900 pC/N,[22] which provides an exceptional opportunity for achieving large change of in-plane uniaxial anisotropy in the FeGaB films, and therefore large tunable FMR frequency range.

The ME coupling in FeGaB/Si/PMN-PT multiferroic heterostructures was investigated at both DC and microwave frequencies by using vibrating sample magnetometer (VSM) and field sweep FMR spectrometer and network analyzer, respectively. A coplanar wave guide (CPW) was used in conjunction with the network analyzer to measure the permeability of the FeGaB/Si/PMN-PT heterostructures, which was schematically shown in Fig. 1.

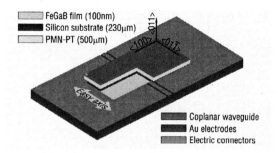

FeGaB film (100nm)
Silicon substrate (230μm)
PMN-PT (500μm)

Coplanar waveguide
Au electrodes
Electric connectors

Fig. 1 Schematic of the sample configuration FeGaB/Si/PMN-PT hetero-structure and permeability measurement setup.

Microwave ME coupling was investigated by measuring the electric field induced changes in permeability spectrum under different static electric fields across PMN-PT substrate. During the measurement, the magnetic easy axis of the FeGaB film was placed parallel to the [100] direction (long axis) of the PMN-PT to minimize the magnetic loss tangent. An external DC magnetic bias field of 20 Oe was applied parallel to the magnetic easy axis of the FeGaB film which was along the long direction [100]. Clearly, the peak FMR frequency change dramatically with external electric field, as shown in figure 2 (a), being 1.1 GHz at -6 kV/cm, 1.6 GHz at zero field, and 2.0 GHz at the coercive field of the PMN-PT single crystal, +2 kV/cm. This equals a tunable frequency of $\Delta f = 900$ MHz, and $\Delta f/[(f_{max}+f_{min})/2] = 58\%$, which is almost one order of magnitude higher than reported results.[20] The electric field induced ferromagnetic frequency change can be interpreted by the Kittel Equation (cgs units):[23]

$$f_{FMR} = \gamma\sqrt{\left(H_k + H_{dc} + \Delta H_{eff}\right)\left(4\pi M_s + H_k + H_{dc} + \Delta H_{eff}\right)},$$

where H_{dc} is the external bias field and H_k is the intrinsic in-plane anisotropy field of the FeGaB film. Besides the FMR frequency change, the initial relative permeability was also changed accordingly, from ~1000 for -6 kV/cm to ~350 for +2 kV/cm, which can be easily understood by the following relation (cgs units):

$$\mu_i = \frac{4\pi M_s}{H_k + H_{dc} + \Delta H_{eff}} + 1.$$

The "buffer-fly" curve of FMR frequencies vs. applied electric field was observed and shown in Fig. 2 (b), which resembled the widely observed piezoelectric strain vs. electric field "butterfly" curves for piezoelectric materials and matched the ferroelectric P-E hysteresis loop of PMN-PT single crystal as well. This once again confirmed that the change of FMR frequency of the FeGaB film results from the ME coupling induced strain in the FeGaB film.

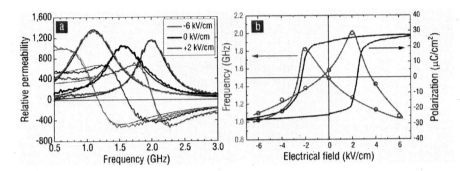

Figure 2. Permeability spectra of the FeGaB/Si/PMN-PT heterostructure under different electric fields (left); The "butterfly" shaped hysteresis of the FMR frequency vs. electric field of the FeGaB/Si/PMN-PT heterostructure, and the ferroelectric P-E hysteresis loop of the PMN-PT single crystal (right).

The static ME couplings in FeGaB/Si/PMN-PT was demonstrated by observing magnetic hysteresis loop changes while applying electric field across PMN-PT substrate as shown in Fig. 3. The remnant magnetization ratio showed a remarkable change from 74% to 16% when the electric field was varied from -6 kV/cm to +2 kV/cm. There is a clear trend of anisotropy field change, in which a electric field of -6 kV/cm corresponded to a minimal anisotropy field while electric field of +2 kV/cm provided a much higher anisotropy field. These anisotropy field changes agreed well with the results from microwave measurements.

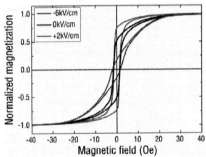

Figure 4. Magnetic hysteresis loops of the FeGaB/Si/PMN-PT heterostructure under different electric fields.

3 FERRITE/FERROELECTRIC HETEROSTRUCTURES DERIVED BY LOW TEMPERATURE SPIN-SPRAY PROCESS

Spin-spray deposition process is a novel technique to plate high crystalline quality spinel ferrite film with different compositions directly from aqueous solution at a temperature less than 90°C.[24] Conventional ferrite film preparation methods such as sputtering, MBE, PLD, etc., all require high temperatures (well above 600°C). In addition, the ferrites can be easily deposited by spin spray processing on various substrates, such as ceramics, glass, metals, etc., making these ferrite films readily integrated onto different integrated circuits. Generally, several chemical reactions are involved in ferrite spin-spray process, which consist of: a) adsorption of Fe^{2+} and other ions M^{n+} (M= Fe, Co, Ni, Zn etc.) on the substrate surface mediated by OH group; b) oxidization process expressed by $Fe^{2+} \xrightarrow{NaNO_2} Fe^{3+} + e$; and c) ferrite film formation accompanying with hydrolytic dissociation which can be expressed by

$$xFe^{2+} + yFe^{3+} + zM^{n+} + 4H_2O \rightarrow (Fe^{2+}, Fe^{3+}, M^{n+})_3O_4 + 8H^+ \ (x + y + z = 1, 2x + 3y + nz = 8).^{[25]}$$

Spin-spray derived ferrite films have been reported with high ferromagnetic resonance frequency up to several GHz, low microwave loss tangents, and high permeability, and have been applied in different RF/microwave magnetic devices.[26-28] New chemical bonds can be formed at the interface between the spin spray deposited ferrite film and ferroelectric substrates during spin-spray processes, leading to strong interface adhesion which is critical for ME coupling.[29] The low synthesis temperature and strong interface bonding provide great opportunities for the spin spray deposited ferrite/ferroelectric heterostructures.

Different spinel ferrites $Ni_{0.23}Fe_{2.77}O_4$ (NFO), $Co_{0.79}Fe_{3.15}O_4$ (CFO), $Ni_{0.26}Zn_{0.1}Fe_{2.63}O_4$ (NZFO), $Zn_{0.1}Fe_{2.9}O_4$ (ZFO) and Fe_3O_4 were spin-spray deposited on ferroelectric substrates at 90°C to form different multiferroic heterostructures. Commercially available PZT, (011) cut single crystal PMN-PT and (011) cut single crystal PZN-PT substrates with a thickness of 0.5 mm were used. An aqueous solution containing M^{2+} ions (M= Fe, Ni, Zn and Co) with different composition was used as ferrite precursor solution. At the same time, an aqueous oxidization solution with 2 mM $NaNO_2$ and 140 mM CH_3COONa and a pH value of 8.0 was used. Through separate nozzles, these two solutions were sprayed at a flow rate of 40ml/min simultaneously onto a spinning hot ferroelectric substrate at 90°C with rotation speed of 150 rpm. N_2 gas was blown into chamber to mitigate the oxidization effects from the oxygen in air. After thirty minutes plating, a uniform ferrite film with a thickness of 1μm was deposited onto ferroelectric substrates with a growth rate of ~30nm/min.

3.1. Characterization of Spin-spray Deposited Ferrite Films

Phase structures of the spin-spray derived ferrite films on ferroelectric substrates were confirmed by the X-ray diffraction. Pure spinel ferrite phases were clearly identified with no obvious preferential crystallographic orientation in all derived ferrite films. Fig. 4(a) showed a typical XRD pattern of spinel ferrite phase of NFO. Strong adhesions between ferrite and ferroelectric phase were observed in all ferrite/ferroelectric heterosturctures. Figure 4(b) showed a high angle annular dark field (HAADF) scanning transmission electron microscopy (STEM) image (Z-contrast) at the NFO/PZT interface, which displayed a uniform ferrite film with a thickness of 1μm on PZT substrate. It is worth noting that the NFO ferrite phase was in favor of growing into the small surface crack of the PZT phase with a crack width of ~100 nm (area I) which was proved by the energy dispersive x-ray (EDX) results. This indicates excellent wetting between the ferrite phase and the phase during the spin-spray deposition process, which leads to a strong adhesion between the two phases and a strong ME coupling.

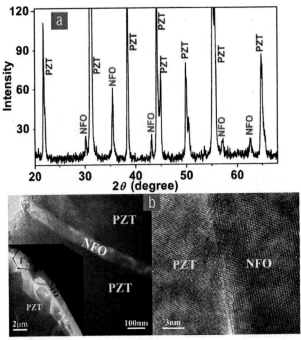

Figure 4. XRD pattern (a) and high resolution TEM image (b) of spin-spray deposited NFO on PZT substrate.

Magnetic properties of spin-spray deposited multiferroic htereostrucutres were investigated by Vibrating Sample Magnetometer (VSM) and field sweep FMR spectrometer at X-band (~9.5GHz) with a TE_{102} cavity. Well-defined hysteresis loops can be observed in all ferrite films with different magnetization and coercivity as shown in Table 1, among which the $Zn_{0.1}Fe_{2.9}O_4$ (ZFO) exhibits the lowest coercivity and FMR linewidth

Table 1. The properties of spin-spray derived spinel ferrites.

Ferrite	Ferroelectrics	$4\pi M_s$	Coercivity (in-plane)	Coercivity (out-plane)	FMR linewidth
$Ni_{0.23}Fe_{2.77}O_4$	PZT	3100	190	260	-
$Co_{0.79}Fe_{3.15}O_4$	PMN-PT	5600	660	740	-
$Ni_{0.26}Zn_{0.1}Fe_{2.63}O_4$	PMN-PT	4500	70	85	350
$Zn_{0.1}Fe_{2.9}O_4$	PMN-PT	5100	36	68	270
Fe_3O_4	PMN-PT	5100	100	114	480
Fe_3O_4	PZT	5100	90	110	590

3.2. Static ME Interactions in Spin-spray Deposited Multiferroic Composites.

Static magnetoelectric coupling of multiferroic heterostructures was studied by electrostatic field induced changes in magnetic hysteresis loops. As shown in Fig. 5(a), the in-plane magnetization process of the Fe_3O_4/PZT multiferroic composite displayed an obvious electric field dependence of the magnetic hysteresis loops while applying electric field across the thickness of PZT. The squareness ratio (SQR or remanence) of the Fe_3O_4/PZT multiferroic composite exhibited a "butterfly" curve dependence on the electric field, with the squareness ratio tunable between 39% ~ 44%. This "butterfly" shape resembled the widely observed piezoelectric strain vs. electric field "butterfly" curves for ferroelectric materials. Similar electric-dependent "butterfly" curves were also observed in the ferromagnetic resonance frequency and magnetization of multiferroic composites,[30] indicating a strain/stress mediated magnetoelectric coupling.

The (011) cut PMN-PT and PZN-PT single-crystal slab has anisotropic in-plane piezoelectric coefficients of d_{31} and d_{32}, which could generate in-plane compressive stress along [100] (d_{31}) and tensile stress along [01-1] (d_{32}) while applying electric field parallel to [011] (d_{33}) direction. The Fe_3O_4 ferrite on PMN-PT displayed a remarkably different magnetization process when it was magnetized along the two in-plane orthogonal directions [01-1] and [100] of the PMN-PT slab, which was reflected by the opposite trend of electric field dependence of the magnetic hysteresis loops as shown in Fig. 5(b). The squareness ratio was minimized to be ~45% at the electric field of +2 kV/cm and maximized to be ~66% at -8 kV/cm when the external magnetic field was applied along the PMN-PT [01-1] (d_{32}) direction. While when the external magnetic field was applied parallel to the PMN-PT [100] (d_{31}) direction, the squareness ratio was maximized to be about 70% at an electric field of +2 kV/cm and minimized to be 40% at -8 kV/cm. It is notable that the anisotropic in-plane electric field induced squareness ratio change from 40% to 70% for the Fe_3O_4/PMN-PT is much larger than the in-plane isotropic squareness ratio change for the Fe_3O_4/PZT, and for Ni-ferrite/PZT. In addition, to further enhance the ME coupling, the (011) cut single crystal PZN-PT with a larger piezoelectric coefficient than that of PMNPT was used as substrate, a large tunable range of the squareness ratio of 40% was achieved in Fe_3O_4/PZN-PT as shown in Fig.5(c), compared to 30% in Fe_3O_4/PMN-PT.

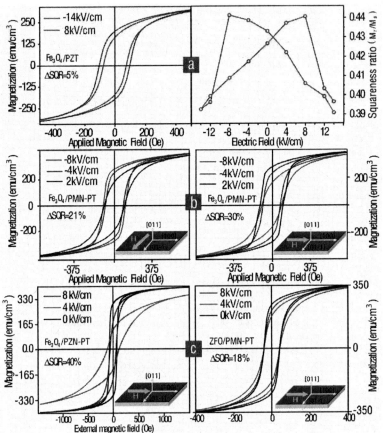

Figure 5. In-plane magnetic hysteresis loops and squareness ratio changes at different E-field in Fe_3O_4/PZT (a); in-plane magnetic hysteresis loops changes under various electric fields with magnetic fields along [100] (d_{31}) and [01-1] (d_{32}) of Fe_3O_4/PMN-PT (b); magnetic hysteresis loop changes with external magnetic fields along [100] (d_{31}) of Fe_3O_4/PZN-PT and ZFO/PMN-PT (c).

3.3. Microwave ME Couplings in Ferrite/ferroelectric Heterostructures.

The ME interactions at microwave frequencies for the spin-spray derived multiferroic heterostructures were demonstrated by an electrostatic induced in-plane magnetic anisotropy field H_{eff} which resulted in large changes in the FMR field at X-band (9.5 GHz). Depending upon whether the H_{eff} is parallel or perpendicular to the applied external magnetic field, the FMR field could be tuned up or down by electric field. The in-plane FMR frequency can be expressed by Kittel equation:

$$f = \gamma\sqrt{(H_r + H_k + H_{eff})(H_r + H_k + H_{eff} + 4\pi M_s)},$$

where γ is the gyromagnetic ratio; H_r is the FMR field that is supplied by an external electromagnet pair; H_k is the in-plane magnetic anisotropic field; $4\pi M_s$ is the magnetization; H_{eff} is the effective in-plane magnetic anisotropy field due to magnetoelectric coupling. The H_{eff} induced by the orthogonal in-plane compressive and tensile stress could be positive or negative, and can be expressed as

$$H_{eff} = \frac{3\lambda_s(\sigma_c - \sigma_t)}{M_s}, \quad \text{in which} \quad \sigma_c = \frac{Y}{1 - v^2}(v d_{32} + d_{31})E \quad \text{and} \quad \sigma_t = \frac{Y}{1 - v^2}(d_{32} + v d_{31})E \quad \text{are the}$$

compressive and tensile stress generated by the electric filed E and derived from the Hooke's law for plane stress condition; Y is the Young's Modulus of zinc ferrite film; v is the Poisson ratio; λ_s is the saturate magnetostriction constant; d_{31} and d_{32} the piezoelectric coefficients along [100] direction and [01-1] direction for (011) cut PMN-PT, respectively. A TE_{102} mode microwave cavity operating at X-band (9.5 GHz) was used to perform the FMR measurements of the ferrite/ferroelectric multiferroic composites. The external bias magnetic field was applied in the ferrite film plane along PMN-PT [100] or PMN-PT [01-1] direction, respectively, with microwave RF field being in-plane and perpendicular to the DC bias field.

Electric field induced ferromagnetic resonance field shifts in different spin-spray deposited multiferroic heterosturctures were shown in Fig.6. A high electrostatically tunable FMR field shift up to 600 Oe, corresponding to a large microwave ME coefficient of 67 Oe cm/kV, was observed in Fe_3O_4/PMN-PT heterostructures when the electric fields across the PMN-PT thickness were changed from 3 kV/cm to -6 kV/cm and when the external magnetic field was applied along the [100] (d_{31}) direction of PMN-PT. In comparison, a giant electrostatically tunable FMR field range of 860 Oe with the linewidth of 330~380 Oe was demonstrated in Fe_3O_4/PZN-PT heterostructure, when the external magnetic fields were applied along in-plane [100] direction (d_{31}: -3000 pC N^{-1}) and the external electric fields were applied across the thickness of PZN-PT substrate starting from 0 kV/cm to 8 kV/cm, corresponding to a ME coupling coefficient of 108 Oe cm /kV. Clearly, compared to Fe_3O_4/PZN-PT, the FMR linewidth was reduced from $\Delta H=480\sim620$ Oe in Fe_3O_4/PMN-PT to $\Delta H=330\sim380$ Oe in Fe_3O_4/PZN-PT, which lead to a significantly enhanced ratio of tunable FMR field over FMR linewidth of 2.5. Static electric field induced FMR field shifts were also investigated in Fe_3O_4/PZT, NZFO/PMN-PT and ZFO/PMN-PT hererostructures, in which FMR field tunable range of 80 Oe, 50Oe and 140 Oe were observed with the FMR linewidth of 500 Oe, 350 Oe and 270 Oe, respectively.

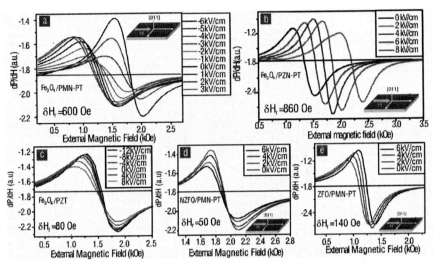

Figure 6. Ferromagnetic resonance absorption spectra of Fe_3O_4/PMN-PT (a), Fe_3O_4/PZN-PT (b), Fe_3O_4/PZT (c), NZFO/PMN-PT (d) and ZFO/PMN-PT (e) while applying different electric fields.

The FMR field of the Fe_3O_4/PZT and Fe_3O_4/PMN-PT multiferroic composites exhibited the characteristic "butterfly" shape in their FMR field vs. electric field curves as shown in Fig. 7, which coincided with the ferroelectric hysteresis loops of the PZT and PMN-PT respectively and were similar to what was observed in the FeGaB/Si/PMN-PT.

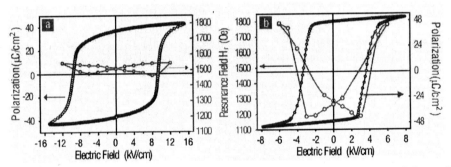

Figure 7. "Butterfly" curves of resonance fields vs. electric fields and ferroelectric hysteresis loops of multiferroic composite Fe_3O_4/PZT (a) and Fe_3O_4/PMN-PT (b).

It can be concluded from Kittel Equation that the FMR frequency in the Fe_3O_4/PMN-PT can be shifted upward or downward, depending on whether the applied magnetic field is parallel or perpendicular to the effective magnetic field induced by the electric field across the (011) cut PMN-PT, which is a natural result of the anisotropic piezoelectric coefficient of the (011) cut PMN-PT. Considering the anisotropic in-plane piezoelectric coefficients of the (011) cut PMN-PT, which can generate orthogonal tensile and compressive stresses under an electric field, a new concept is introduced that FMR frequency could be tuned up or down and achieve a larger tunable range by applying the external magnetic field parallel to [100] (d_{31}) or [01-1] (d_{32}) direction of the (011) cut PMN-PT. This idea was demonstrated in the Fe_3O_4/PMN-PT multiferroic composite. As shown in Fig. 8, an external magnetic field applied along the [01-1] direction led to a reduced FMR field of 1200 Oe (Fig. 8(a)). When the external field was parallel to the [100] direction, the FMR fields were shifted up to 2200 Oe. The total resonance field shift was δH_r=1000 Oe when the external magnetic field was applied to be parallel to [100] and [01-1] direction. These tuning up or down of FMR fields by changing applied magnetic field direction also were occurred in Fe_3O_4/PZN-PT due to the anisotropic piezoelectric effects as shown in Fig.8(b). The total tunable range was up to 1450 Oe (Fig.8 (c)). This constitutes a simple but effective approach for achieving twice the tunable FMR frequency range, leading to new opportunities of magnetostatically tunable magnetic device design with large effective magnetic field tunability.

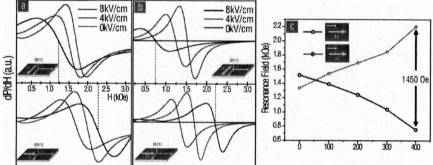

Figure 8. FMR absorption spectra while the external mangetic field along tensional [01-1] and compressive [100] directions of Fe_3O_4/PMN-PT (a), Fe_3O_4/PZN-PT (b), and the total magnetic field shifts is up to 1450 Oe when magnetic field is applied along [01-1] and [100] of the Fe_3O_4/PMN-PT (c).

4. CONCLUSION

In summary, two kinds of multiferroic heterostructures were made with metallic FeGaB film and spin-spray deposited ferrite films on different ferroelectric substrates, which show giant magnetoelectric coupling. A summary table of the magnetoelectric coupling of some multiferroic heterostructures is made for comparison as shown in table 2. A large electric field tunable FMR frequency of 900 MHz was observed in FeGaB/Si/PMN-PT heterostructure; while a giant electrostatically tunable FMR field range of 860 Oe was demonstrated in Fe_3O_4/PZN-PT heterostructure, corresponding to a ME coefficient of 108 Oe cm kV^{-1}. In addition the FMR linewdith was reduced from 480~620 Oe in Fe_3O_4/PMN-PT to 330~380 Oe in Fe_3O_4/PZN-PT, leading to a ratio of tunable FMR field over FMR linewidth of 2.5. The giant electrostatically tunable FMR frequency and FMR field and low synthesis temperature make these multiferroic heterostructures great candidates for applications in electrostatically tunable microwave multiferroic devices.

Talbe 2. The comparison of some microwave ME coupling results in different multiferroic heterosturcutres. (YIG: yttrium iron garnet ferrite; GG: gadolinium gallium garnet)

Structure	δH (Oe)	δH/ δE (Oe cm/kV)	δf (MHz)	δf /δE (MHz cm/kV)	References
YIG/PZT			40	18	31
YIG/PZT	--	--	110	36.7	20
YIG/GGG/PMN-PT	46	5.75	--	--	3
Ni2MnGa/PMN-PT	230	41	--	--	32
FeCoB/Glass/PZT	--	--	50	3.1	33
FeGaB/Si/PZT	--	--	110	6.9	33
FeGaB/Si/PMN-PT	30	3.75	900	112.5	Present
FeGaB/PMN-PT	330	33	--	--	To be published
FeGaB/PZN-PT	750	94	5820	970	To be published
NZFO/PMN-PT	50	8	--	--	Present
ZFO/PMN-PT	140	23	--	--	Present
Fe3O4/PMN-PT	600	67	--	--	Present
Fe3O4/PZN-PT	860	107.5	--	--	Present

ACKNOWLEDGEMENT

This work is sponsored by NSF and ONR.

REFERENCE

[1] J. F.Scott, Data storage: Multiferroic memories, *Nature Materials*, **6**, 256, (2007).

[2] M. Fiebig, Revival of the magnetoelectric effect, *J. Phys. D*, **38**, R123, (2005).

[3] S. Shastry, G. Srinivasan, M. I. Bichurin, V. M. Petrov, A. S. Tatarenko, Microwave magnetoelectric effects in single crystal bilayers of yttrium iron garnet and lead magnesium niobate-lead titanate, *Phys. Rev. B*, **70**, 064416, (2004).

[4] C. W. Nan, M. I. Bichurin, S. X. Dong, and D. Viehland, Multiferroic magnetoelectric composites: Historical perspective, status, and future directions, *J. Appl. Phys.*, **103**, 031101, (2008).

[5] J. Zhai, Z. Xing, S. X. Dong, J. F. Li, D. Viehland, Magnetoelectric Laminate Composites: An Overview, *J. Am. Ceram. Soc.*, **91**, 351, (2008).

[6] W. Eerenstein, N. D. Mathur, J. F. Scott, Multiferroic and magnetoelectric materials, *Nature*, **442**, 759, (2006).

[7] T. Lottermoser, T. Lonkai, U. Amann, D. Hohlwein, J. Ihringer, M. Fiebig, Magnetic phase control by an electric field, *Nature*, **430**, 541, (2004).

[8] S-W. Cheong, M. Mostovoy, Multiferroics: a magnetic twist for Ferroelectricity, *Nature Materials*, **6**, 13, (2007).

[9] H. Zheng, J. Wang, S. E. Lofland, Z. Ma, L. Mohaddes-Ardabili, T. Zhao, L. Salamanca-Riba, S. R. Shinde, S. B. Ogale, F. Bai, D. Viehland, Y. Jia, D. G. Schlom, M. Wuttig, A. Roytburd, R. Ramesh, Multiferroic BaTiO$_3$-CoFe$_2$O$_4$Nanostructures, *Science*, **303**, 661, (2004).

[10] A. Brandlmaier, S. Geprägs, M. Weiler, A. Boger, M. Opel, In situ manipulation of magnetic anisotropy in magnetite thin films, *Phys. Rev. B*, **77**, 104445, (2008).

[11] M. Liu, X. Li, J. Lou, S. Zheng, K. Du, Nian X. Sun, A modified sol-gel process for multiferroic nanocomposite films, *J. Appl. Phys.*, **102**, 083911, (2007).

[12] J. Ma, Z. Shi, and C. W. Nan, Magnetoelectric Properties of Composites of Single Pb(Zr,Ti)O3 Rods and Terfenol-D/Epoxy with a Single-Period of 1-3-Type Structure, *Adv. Mater.*, **19**, 2571, (2007).

[13] Petrov V M, Srinivasan G, Bichurin M I and Gupta A, Theory of magnetoelectric effects in ferrite piezoelectric nanocomposites, Phy. Rev. *B*, **75**, 224407, (2007).

[14] M. Liu, X. Li, H. Imrane, Y. Chen, T. Goodrich, K. S. Ziemer, J. Y. Huang, and N. X. Sun, Synthesis of Ordered Arrays of Multiferroic NiFe$_2$O$_4$–Pb(Zr$_{0.52}$Ti$_{0.48}$)O$_3$ Core-Shell Nanowires, *Appl. Phys. Lett.*, **90**, 152501, (2007).

[15] J. Lou, D. Reed, M. Liu, N. X. Sun, "Electrostatically tunable inductors with multiferroic composites," *Appl. Phys. Lett.* Submitted

[16] J. Zhai, Z. Xing, S. X. Dong, J. F. Li, and D. Viehland, Detection of pico-Tesla magnetic fields using magneto-electric sensors at room temperature, *Appl. Phys. Lett.*, 88, 062510, (2006).

[17] Y. K. Fetisov, and G. Srinivasana, Electric field tuning characteristics of a ferrite-piezoelectric microwave resonator, *Appl. Phys. Lett.* **88**, 143503, (2006).

[18] A. Ustinov, G. Srinivasan, and B. A. Kalinikos, Ferrite-ferroelectric hybrid wave phase shifters, *Appl. Phys. Lett.*, **90**, 031913, (2007).

[19] C. Pettiford, S. Dasgupta, J. Lou, S. D. Yoon, and N. X. Sun, "Bias Field Effects on Microwave Frequency Behavior of PZT/YIG Magnetoelectric Bilayer," IEEE Trans. Magn. **43**, 3343-3345, (2007).

[20] A. S. Tatarenko, V. Gheevarughese, G. Srinivasan, "Magnetoelectric microwave bandpass filter," *Electro. Lett.*, **42**, 540, (2006).

[21] J. Lou, R. E. Insignares, Z. Cai, K. S. Ziemer, M. Liu, and N. X. Sun, "Soft magnetism, magnetostriction, and microwave properties of FeGaB thin films," *Appl. Phys. Lett.*, **91**, 018254, (2007).

[22] P. Han, W. Yan, J. Tian, X. Huang, and H. Pan, "Cut directions for the optimization of piezoelectric coefficients of lead magnesium niobate–lead titanate ferroelectric crystals", *Appl. Phys. Lett.*, **86**, 052902, (2005).

[23] C. Kittel, "On the Theory of Ferromagnetic Resonance Absorption," *Phys. Rev.* vol. **73**, 155, (1948).

[24] M. Abe, Y. Tamaura, Ferrite-plating in aqueous solution: a new method for preparing magnetic thin film, *Jpn. J. Appl. Phys.*, **22**, L511, (1983).

[25] M. Abe, Ferrite plating: a chemical method preparing oxide magnetic films at 24–100°C, and its applications, *Electrochimica Acta*, **45**, 3337, (2000).

[26] M. Taheri, E. E. Carpenter, V. Cestone, M.M. Miller, M.P. Raphael, M.E. McHenry, and V.G. Harris, Magnetism and structure of ZnxFe3–xO4 films processed via spin-spray deposition, *J. Appl. Phys.* **91**, 7595, (2002).

[27] K. Kondo, T. Chiba, H. Ono, and S. Yoshida, Y. Shimada, N. Matsushita, M. Abe, High-frequency transport properties of spin-spray plated Ni–Zn ferrite thin films, *J. Appl. Phys.*, **93**, 7127, (2003).

[28] K, Kondo, S, Yoshida, and H, Ono, M. Abe, Spin sprayed Ni(–Zn)–Co ferrite films with natural resonance frequency exceeding 3 GHz, *J. Appl. Phys.*, **101**, 09M502, (2007).

[29] M. Liu, O. Obi, J. Lou, S. Stoute, J. Y. Huang, Z. Cai, K. S. Ziemer, N. X. Sun, Spin-spray deposited multiferroic composite $Ni_{0.23}Fe_{2.77}O_4$/Pb(Zr,Ti)O$_3$ with strong interface adhesion, *Appl. Phys. Lett.*, **92**, 152504, (2008).

[30] C. Thiele, K. Dörr, O. Bilani, J. Rödel, L. Schultz, Influence of strain on the magnetization and magnetoelectric effect in La0.7A0.3MnO3/PMN-PT(001) (A=Sr,Ca), Phys. Rev. B, **75**, 054408, (2007).

[31] Y. K. Fetisov and G. Srinivasan, Nonlinear electric field tuning characteristics of yttrium iron garnet–lead zirconate titanate microwave resonators, *Appl. Phys. Lett.*, **93**, 033508 (2008).

[32] Y. Chen, J. Wang, M. Liu, J. Lou, N. X. Sun, C. Vittoria, and V. G. Harris, Large electric field-induced magnetic field tunability in a laminated Ni2MnGa/PMN-PT multiferroic heterostructure", *Appl. Phys. Lett.* **93**, 112502 (2008).

[33] C. Pettiford, J. Lou, L. Russell, N. X. Sun, Strong Magnetoelectric Coupling at Microwave Frequencies in Metallic Magnetic Film / Lead Zirconate Titanate Multiferroic Composites, *Appl. Phys. Lett.*, **92**, 122506 (2008).

MECHANICAL STRAIN AND PIEZOELECTRIC PROPERTIES OF PZT STACKS RELATED TO SEMI-BIPOLAR ELECTRIC CYCLIC FATIGUE

Hong Wang[*], Hua-Tay Lin, Thomas A. Cooper, and Andrew A. Wereszczak
Ceramic Science and Technology Group
Materials Science and Technology Division
Oak Ridge National Laboratory[†]
Oak Ridge, TN 37831

ABSTRACT

PZT stacks that have an inter-digital internal electrode configuration were tested to more than 10^8 cycles. A 100-Hz semi-bipolar sine wave with a field range of +4.5/-0.9 kV/mm was used in cycling with a concurrently applied 20 MPa preload. Significant reductions in mechanical strain and piezoelectric coefficients were observed during the cycling depending on the measuring condition. Extensive surface discharges were also observed. These surface events, as well as related breakdown, resulted in the erosion of external electrode and the outcrop of internal electrode, which partially accounted for the observed reductions. The results from this study demonstrate the feasibility of using a semi-bipolar mode to drive a PZT stack under a mechanical preload and illustrate the potential fatigue performance of the stack in service.

INTRODUCTION

The engine industry is pursuing the use of bipolar electric loading to enhance the mechanical actuation of piezo or PZT stacks or actuators [1] for fuel injection. Theoretically, a reverse electric field at a limited level can introduce beneficial depolarization and increase the number of switchable domains. Hence, a greater piezoelectric response of a stack can be promoted. This is mainly based on tests of single-layer PZT [2] and limited work on several commercial PZT stacks [3]. However, the fatigue performance of PZT stacks under the bipolar electric loading mode, specifically semi-bipolar mode, is essentially uncharacterized.

This study focuses on commercial stacks of PZT-5A that have a Curie temperature of 350°C and considered as a good candidate for the fuel injection system of heavy duty diesel engines. These stacks were tested to a specific number of cycles with a semi-bipolar sine wave under a mechanical preload. The fatigue responses of the stacks in terms of mechanical strain and piezoelectric coefficients are presented and discussed.

EXPERIMENTAL TECHNIQUE AND ANALYTICAL APPROACH

Mechanical and Electric Loading

A mechanical loading test system was developed for this study's fatigue testing [4]. The loading frame had a dead weight (or "preloading") capability. The arrangement of the piezo stack section located between a loading rod and base is shown in Fig. 1. A load cell (Sensotec 31, Columbus, OH) mounted within the load train was used to continually

---

[*] Corresponding author. Tel: 1-865-574-5601; email: wangh@ornl.gov
[†] Managed by UT-Battelle, LLC, for the U.S. Department of Energy under contract DE-AC05-00OR22725.

measure and monitor compressive force. The piezo stack under test was positioned between two steel beams, and those functioned as the supporters for the sensor and the measuring target, respectively. Capacitance displacement gages (Capacitec HPT-40, Ayer, MA) located on each side of the stack were used to continuously measure displacement (two gages were used to enable averaging).

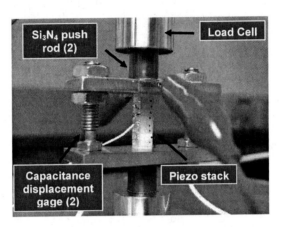

Fig. 1. Set-up for the piezo stack section of a piezo stack fatigue test facility.

A LabView program was used to generate the sine waveform throughout this study. A high-voltage (HV) amplifier (Trek PZD350M/S, Medina, NY) having a current capacity of ±400 mA was used in the fatigue cycling. Another HV amplifier (Trek 609E-6, Medina, NY) having a lower current capacity (±20 mA) but higher voltage output (±4 kV) was employed in measurements to evaluate the performance of PZT stack.

PZT Stack

Each PZT stack (Noliac, Denmark) had an overall size of 5mm x 5mm x 18mm and was composed of 8 active plates (modules) and 2 inactive endplates. Each individual active plate (5mm x 5mm x 2mm) was a multilayer actuator with an inter-digital electrode (IDE) configuration that included 30 PZT layers each 67 μm thick. The 5mm x 2mm surfaces of the active plates were coated with silver that served as an external electrode. These silver-coated plates were interconnected with a bus wire in a zigzag pattern. The rated voltage and capacitance of PZT stack were 200 V and 0.76 μF, respectively.

Data Acquisition and Test Procedure

The data acquisition system included amplifiers, a data connection board (NI SCB-68, Austin, TX), a data acquisition card (NI USB-6251, Austin, TX), and a

LabView program. The LabView program both drove the testing and collected measurement signals.

Three electric waveforms were used throughout the testing. A 100-Hz, +300/-60 V sine waveform (semi-bipolar, 4.5/-0.9 kV/mm) was used with a 20 MPa preload for the accumulation of cycles. Two 10-Hz waveforms were used to evaluate performance changes at a specified cycle number that involved smaller positive peak-values, +200/0 V (unipolar, +3.0/ 0 kV/mm) and +200/-60 V (semi-bipolar, +3.0/-0.9 kV/mm), and successive preloads of 0.7, 20, 0.7 MPa. The 0.7 MPa preload, the minimum allowed load to stabilize the stack, was used in repoling the specimen. For each measurement, the data were sampled at each second of the first 10s period. A data truncation method as reported by Wang et. al. [4] was used. To eliminate any uncertainty due to the cycling-induced temperature, measurements were taken 1 h after the cycling had been paused. Results will be reported for two PZT stacks (No. 02 and 05) in this study.

Analytical Approach

A typical measurement consisted of capturing signals from six channels: one input voltage (from function generator), two capacitance gages, one load cell, and two responses (voltage and current) from the HV amplifier. The raw signals were then processed to determine relevant quantities including the electric field (E) and mechanical strain (S) [4].

The peak field-induced response, particularly the mechanical strain S_p, was used to quantify the response of each tested PZT stack. Meanwhile, the piezoelectric hysteresis related to a S-E loop was characterized as

$$U_d = -\oint SdE .$$ (1)

This definition was introduced by Hall [5] and its application to PZT stacks was also demonstrated by Wang et. al. [4]. Fourier Transformations were used to decompose acquired signals in this study. These decomposed signals served as inputs to determine the frequency response function (FRF) of the stack being tested [4]. The obtained FRF of strain with respect to electric field defines a complex piezoelectric coefficient d^* as:

$$d^* = d' - jd'' = de^{-j\delta_p} ,$$ (2)

where δ_p is the phase delay or loss angle whose tangent defines the dissipation factor. Similar approaches were also used by others [5,6,7] in characterizing responses of PZT stacks.

EXPERIMENTAL RESULTS

Prefatigue Results

The manufacturer's reported nominal free stroke of the stacks was 23.6 μm under the rated voltage of 200 V or a strain of 1,475 με at 3 kV/mm. The prefatigue test on a stack resulted in a higher value (1,600 $\mu\varepsilon$) than that quoted though. This difference could be attributed to batch-to-batch variance in manufacturing of stacks as can be seen from the 10-15% tolerance. Meanwhile, it was found that the hysteresis (850 kV/mm) of the stack in this study was much larger than that (210 kV/mm) of a PSI actuator [4]; apparently, the involved PZT materials, internal electric configuration and measurement conditions contributed to the observed discrepancy.

Our results demonstrated that the preload effect on the mechanical displacement was not as significant as that on the PSI actuator [4] over the preload range from 0.7 to 60 MPa. Nevertheless, the strain exhibited a maximum value around 20 MPa. The level of preload at the maximum strain agreed well with that in a static measurement by Andersen et. al. [8], so that preload level was used in the cyclic fatigue. As expected, the semi-bipolar (+3.0/-0.9 kV/mm at 10Hz) boosted both the mechanical strain (from 1,600 to 2,300 $\mu\varepsilon$) and charge density (from 0.05 to 0.08 C/m^2). In fact, the effects of a reverse field in semi-bipolar loading on the strain and piezoelectric hysteresis were shown to be more significant than those of a mechanical preload within the tested range. The discussion in this study will focus on measurements under only one preload, the 20-MPa case.

The prefatigue level of piezoelectric coefficient was evaluated to be around 450 - 520 pm/V in the case of the stabilizing preload (0.7 MPa), which is actually close to that reported by Andersen et. al. [6].

Fatigue Results

Mechanical strain

After 10^8 cycles, both the strain and piezoelectric hysteresis of No. 02 were reduced (Fig. 2). A more pronounced reduction was exhibited in the measurement with semi-bipolar loading than with unipolar loading. The strain exhibited a monotonic decrease in the measurement with a bipolar mode along with the accumulation of cycles, whereas that in a unipolar mode showed a restoration at around 10^6 cycles. The restoration in unipolar loading was also observed in the fatigue of several other PZT stacks [4, 9, 10]. The variation of the piezoelectric hysteresis during the cyclic fatigue appeared to be quite similar to that of the strain.

Stack No. 05 exhibited different variation patterns in the strain and piezoelectric hysteresis (Fig. 2), though it was tested under the same condition as that of No. 02. A fast drop was observed at around 10^6 cycles in both quantities, corresponding to a stage when extensive surface discharges appeared on electrode surfaces of the stack. No. 05 had higher prefatigue levels of the strain and piezoelectric hysteresis than No. 02, but lower post-fatigue levels as a consequence of cycling.

Curves of normalized quantities versus cycle number describe more clearly the trend associated with a tested stack (Fig. 3). More than 50% decreases in the strain and piezoelectric hysteresis were observed for No. 05, and about 25% decreases were seen for No. 02. For No. 05, the varying degrees of the strain or piezoelectric hysteresis were similar under both loading modes; for No. 02, those were somehow different, especially that of the piezoelectric hysteresis. Overall, the degree of reduction in mechanical strain induced by a semi-bipolar fatigue in these stacks was considerably higher than that caused by a unipolar fatigue (~ 6%) [4,10].

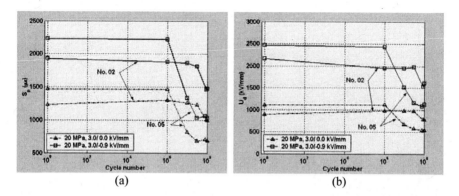

(a)   (b)

Fig. 2. Variations of peak field-induced strain (a) and piezoelectric hysteresis (b) as a function of the number of cycles.

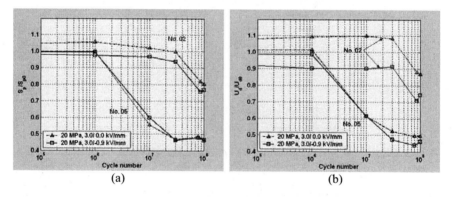

(a)   (b)

Fig. 3. Variations of normalized peak field-induced strain (a) and normalized piezoelectric hysteresis (b) as a function of the number of cycles (10^5- 10^8).

Piezoelectric coefficients

The value of the first harmonic appeared to be relatively low compared to that of the dc component [Figs. 4(a) and 4(b)] indicating the piezoelectric coefficient depended on the driving frequency. The amount of enhancement in piezoelectric coefficient by a reverse field also depended on the frequency. And that can be seen from a larger increase in the dc component than in the first harmonic in the measurement of a semi-bipolar mode. Variation patterns of the dc and the first harmonic are quite similar to that of the strain [Figs. 2(a), 4(a) and 4(b)]. For example, the fast drop in the strain of No. 05 is reflected in the fatigue curves of its dc and first harmonic components. In addition, the degree of piezoelectric enhancement by the reverse field was decreased with the accumulation of cycles; this is evidenced from the shrinking distance between the two curves of a tested stack, especially those of the dc component.

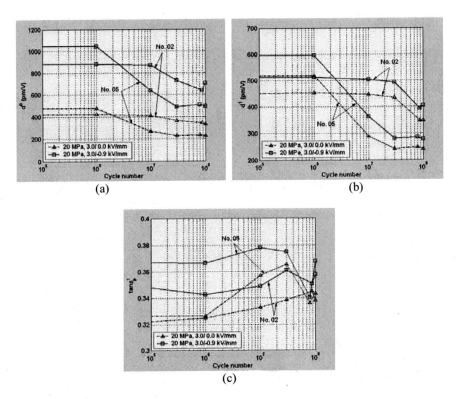

Fig. 4. Variations of (a) the dc and (b) the first harmonic of field-induced strain and (c) piezoelectric loss tangent as a function of the number of cycles (10^5- 10^8).

Use of a reverse field in semi-bipolar mode resulted in a larger piezoelectric loss tangent than in unipolar mode. As a result, the variation of the piezoelectric loss tangent with the cyclic fatigue was related to the load mode of measurement too. Particularly, these loss tangents exhibited approximately decreasing and increasing trends under semi-bipolar and unipolar modes, respectively. And this is a little different from the behavior of the piezoelectric hysteresis that dropped toward 10^8 cycles in both load modes [Figs. 2(b) and 4(c)]. In the later stage of fatigue, the variation also featured a sizable fluctuation.

The normalized dc and first harmonic components [Figs. 5(a) and 5(b)] exhibit similar variation patterns to that of the normalized strain [Fig. 3(a)]. The shifted piezoelectric loss tangent [Fig. 5(c)] shows a decreasing tendency followed by some fluctuation in a semi-bipolar measurement, and an increase trend in a unipolar measurement.

An additional fatigue session of 0.85×10^8 cycles was conducted for No. 05 after 10^8 cycles to explore any tendency of deterioration. No significant variation was observed in the piezoelectric coefficients, but a considerable increase in the piezoelectric loss tangent was obtained.

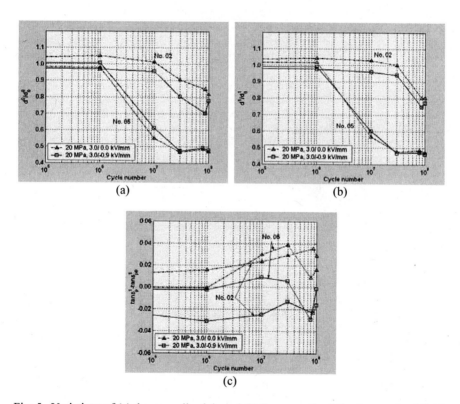

Fig. 5. Variations of (a) the normalized dc and (b) the normalized first harmonic of field-induced strain and (c) shifted piezoelectric loss tangent as a function of the number of cycles (10^5- 10^8).

Deterioration of surface

Both stacks exhibited surface damages after more than 10^8 cycles. But No. 5 showed more, corresponding to larger reductions in mechanical strain and piezoelectric coefficients. The effect of fatigue on the stack surface can be seen from deteriorations of the interconnect joints and the inter-plate bonds. These areas became increasingly darkened because of frequent nearby discharges during the fatigue testing. Within one module of No. 02, a thin plate-through fracture was also found near an internal electrode, arising perhaps from an interface delamination between the affected PZT layer and the electrode.

Figure 6 presents a set of optical micrographs obtained on surfaces of No. 05 after an extended fatigue testing (1.85×10^8 cycles). On the positive electrode side [Fig. 6(a)], dark areas were observed on the plate/plate interfaces and five consecutive solder joints. The sizable dark areas occurred on both sides of the affected part of the interface; the boundary to which the nearby silver coating was removed was quite discernable as shown in Fig. 6(b). An arch-opening zone was usually developed around the foot of the solder joint where the silver coating eroded and the underlying PZT layer emerged. Over the opening zone, an array of equally-spaced gaps was often found whose spacing was estimated to be approximately 134 μm [Fig. 6(c)]. Because this was twice the thickness of a single PZT layer, these linear gaps have been identified to be the outcrops of (positive) internal electrodes as a result of deep erosion. In some local areas, the regular striations from the coating process were replaced by a rough or grainy surface with residuals from surface discharges that featured bubbles of various sizes.

The silver coating was about 150 to 210 μm thick as can be seen from the lateral surface [Figs. 6(d)]. It did not cover the electrode face (5mm x 2mm) of the plate evenly. A dark spot caused by discharges took place at a junction between two external electrode coatings, though the local positive electrode surface did not show any appreciable sign of darkening [Figs. 6(a)].

On the negative electrode side, dark areas occurred more intensively on three separate solder joints. These dark areas were located near the bottoms of the solder joints and the neighboring plate/plate interfaces. An opening zone was also revealed around the bottom of the affected solder joint.

DISCUSSION

It has been demonstrated that, if the strain reduction of a PZT stack is less than 10%, the degradation of performance is mostly due to the decreased domain activities arising from the pinning of non-$180°$ domain and domain wall; and the charged point-defect is mainly responsible for the pinning [2,4,10]. Because of more than 50% reduction observed in this study, it is believed that the micro flaws, micro cracks and surface discharges were activated mechanisms in deteriorating the performance of the stack in the cycling process [2,4]. As the microstructure analysis on tested stacks is undergoing, the following will examine the related mechanism to the surface discharges.

Under a high field of loading, the cycling can inject electric charges into the (external) electrode. When the electric field near the surface was high enough, the breakdown or partial discharge would occur in the surrounding air instead of within the insulation. The partial discharge not only dissipated substantial energy into the surrounding air, but also deteriorated the PZT stack. Both the chemical reactions and physical attacks from partial discharge-induced particles contributed to the surface degradation of the stack. Moreover, the local high temperature induced by discharges can facilitate the erosion of silver coating relevant to the chemical reactions. This does not exclude the direct melting of coating because the melting temperature of silver (961.78°C) could be readily reached. The exposed PZT layer also collected deposits from partial discharges that could serve as a point of field enhancement and dynamically drive the partial discharge. An electric tree could be initiated [11], and the induced structural change would finally damage the insulation system.

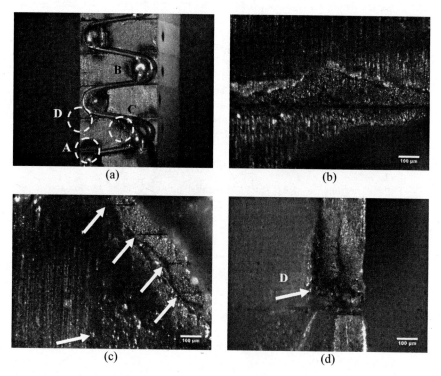

Fig. 6. Micrographs of No. 05 for (a) the positive electrode side showing various deteriorated parts: plate interface segment A, arch openings B & C, and external electrode junction D, (b) enlarged plate/plate interface segment A, (c) enlarged opening area C showing an array of linear gaps and a local rough surface with bubbles, (d) enlarged dark area on lateral surface near the junction D.

Table 1 gives a survey on the surface deteriorations in various scenarios. It is shown that the total number of the deterioration events is correlated to the strain reduction.

Table 1 Surface deterioration events based on optical microscopy*

Stack	SR (%)	Positive electrode side			Negative electrode side			Total
		SJ	IPB	ILE	SJ	IPB	ILE	
No. 02	24	2		2			1	5
No. 05	54	4	10		3	3		20

*SR- strain reduction; SJ – solder joint darkened; IPB – inter-plate bond segment darkened; ILE –inter-layer (internal) electrode darkened.

CONCLUSION

Cyclic fatigue tests were conducted on commercial PZT-5A stacks using a semi-bipolar 100-Hz sine wave that had an equivalent field range of +4.5/-0.9 kV/mm and a 20 MPa preload. More than 10^8 cycles were accumulated and the performance of tested stacks was measured and evaluated. The following conclusions can be drawn:

1. Mechanical strain and piezoelectric coefficient of tested PZT stacks were reduced 24 to 54%.
2. Fatigue response of PZT stacks depended on the loading mode, and a semi-bipolar load accelerated the fatigue.
3. Surface discharges partially accounted for the reductions observed during the cyclic fatigue.
4. Pinned domains and domain walls, micro-cracks, and local breakdown contributed to the reductions also.

ACKNOWLEDGMENTS

This research was sponsored by the U.S. DOE, Office of Vehicle Technologies, as a part of the Propulsion Materials Program, under contract DE-AC05-00OR22725 with UT-Battelle, LLC. The authors thank Drs. Michael J. Lance and Amit Shyam for reviewing the manuscript.

REFERENCES

[1] Mark C. Sellnau, and P. J. Dingle, Private communications, 2007.
[2] D. C. Lupascu, and J. Rodel, Adv. Eng. Mater., 7, 882 (2005).
[3] M. Mitrovic, G. P. Carman, and F. K. Straub, Int. J. Solids and Struct., 38, 4357 (2001).
[4] H. Wang, A. A. Wereszczak, and H.-T. Lin, J. Appl. Phys., 105, 014112 (2009).
[5] D. A. Hall, P-E measurement and analysis software, in NPL Report CMMT (A) 98, Electric property measurements for piezoelectric ceramics: Technical notes, M. G. Cain et. al., March, 1998, p. 70.
[6] B. Andersen, E. Ringgarard, T. Bove, A. Albareda, and R. Perez, Performance of piezoelectric ceramic multiplayer components based on hard and soft PZT, Actuator

2000, 7th International Conference on New Actuators, 19-21 June 2000, Bremen, Germany.

[7] A. Heinzmann, E. Hennig, B. Kolle, D. Kopsch, S. Richter, H. Schwotzer, and E. Wehrsdorfer, Properties of PZT multilayer actuators, Actuator 2002.

[8] B. Andersen, E. Ringgaard, and L. S. Nielsen, Static and dynamic performance of stacked multilayer actuators based on hard and soft PZT, Actuator 2000, 7th International Conference on New Actuators, 19-21 June 2000, Bremen, Germany.

[9] D. Wang, Y. Fotinich, and G. P. Carman, J. Appl. Phys., 83, 5342 (1998).

[10] P. M. Chaplya, M. Mitrovic, G. P. Carman, and F. K. Straub, J. Appl. Phys., 100, 124111 (2006).

[11] K. N. Mathes, Surface failure measurements, in Engineering Dielectrics, Vol. IIB, Electrical Properties of Solid Insulating Materials: Measurement Techniques, ASTM Special Technical Publication 926, Ed. R. Bartnikas, 1987, p. 221.

INTERGRANULAR FRACTAL IMPEDANCE ANALYSIS OF MICROSTRUCTURE AND ELECTRICAL PROPERTIES OF RARE-EARTH DOPED BaTiO3

V.Mitic[1,2], V.B.Pavlovic[3], Lj.Kocic[1], V.Paunovic[1], D.Mancic[1]

[1]Faculty of Electronic Engineering, University of Nis, Nis, Serbia

[2]Institute of Technical Sciences of SASA, Belgrade, Serbia

[3]Faculty of Agriculture, University of Belgrade, Belgrade, Serbia

ABSTRACT

Microstructure properties of barium-titanate based materials, expressed in grain boundary contacts, are of basic importance for electric properties of these material. In this study, the model of intergranular impedance applied on two-grain contact is considered. Globally, BaTiO3-ceramics sample is consisted of a huge number of mutually contacted grains which form clusters. Such clusters can be presented as a specific fractal formations. For each of them, it is possible to establish the equivalent electrical model and, for defined set of input parameters, using symbolic analysis, obtain the frequency diagram. The influence of fractal structure is especially stressed. Realizing the totality of relations between clusters grains groups, their microelectrical schemes and corresponding frequency characteristics, from one side, and global equivalent electrical scheme and corresponding acquired frequency characteristics of BaTiO3-ceramics samples, on the other side, we set a goal of coinciding experimental results with the summing effect of microelectric equivalent schemes. The model is successfully tested on doped barium- titanate ceramics.

INTRODUCTION

Doped barium-titanate ceramic is attracting much interest for its application as resistors with a positive temperature coefficient of resistivity (PTCR), multilayer ceramic capacitors (MLCC), thermal sensors etc [1,2]. In the process of BaTiO3-ceramics consolidation, technological parameters like pressing pressure, initial sample's density, sintering temperature and time, as well as different dopants essentially determine final electrical properties of the ceramics. A slight change of particular consolidation parameter, or the change of dopant's concentration can significantly change the microstructure, thus influencing electrical properties of the speciments. Since grain size and distribution considerably affect electrical properties of barium-titanate based materials, correlation of their microstructure and electrical properties has been investigated most extensively by numerous authors [3-5]. It has been shown that electrical properties of undoped and doped BaTiO$_3$-ceramics are mainly controlled by barrier structure, domain motion of domain boundaries and the effects of internal stress in the grains [6-7]. Therefor, microstructure properties of barium-titanate based materials, expressed in grain boundary contacts, are of basic importance for electric properties of these material. Ussually, for electronic materials design, the microstructure of this type of ceramics can be considered to be as sketched in the cross section of Fig. l(a) and simplified by a brick wall model [8] (Fig. l(b)).

The model illustrates the polarization and the conduction contributions to the total impedance of the system which can be represented by an electrical equivalent network consisting of three RC branches. The elements of the network may be attributed to the permittivities, conductivities, and geometrical extensions of the bulk of the grains (G), the grain boundary regions (GB), and the electrical interfaces (EI). It is well known that both intergranular structure and electrical properties depend on ceramics diffusion process. Therefor it is essential to have an equivalent cirquit model that provides a realistic represantation of the electrical properties. Having this in mind, the purpose of this article is to determine an intergranular impedance model based on intergranular capacity of BaTiO$_3$-ceramics doped with different rare earth of additives.

EXPERIMENTAL PROCEDURE

The samples were prepared from high purity (>99,98) commercial BaTiO$_3$ powder (MURATA) with [Ba]/[Ti]=1,005 and reagent grade Er$_2$O$_3$ and Yb$_2$O$_3$ powders (Fluka chemika). Er$_2$O$_3$ and Yb$_2$O$_3$ dopants were used in the amount to have 0.01, 0.1 and 0.5 wt% Er or Yb in BaTiO$_3$. Starting powders were ball milled in ethyl alcohol for 24 h using polypropylene bottle and zirconia balls. After drying at 200°C for several hours, the powders were pressed into disk of 7 mm in diameter and 3 mm in thickness under 120 MPa. The compacts were sintered 4 h in air up to 1380°C. The microstructures of as sintered or chemically etched samples were observed by scanning electron microscope (JEOL-JSM 5300) equipped with energy dispersive x-ray analysis spectrometer (EDS-QX 2000S system). Prior to electrical measurements silver paste was applied on flat surfaces of specimens. Capacitance, dissipation factor and impedance measurements were done using Agilent 4284A precision LCR meter in the frequency range of 20 to 10^6 Hz. The illustrations of the microstructure simulation, were generated by Mathematica 6.0 software.

RESULTS AND DISSCUSSION

In order to establish the model of intergranular impedance for doped barium titanate, it is important to notice that microstructure properties of BaTiO$_3$ based materials, expressed in their grain boundary contacts, are of basic importance for electric properties of these materials. The barrier character of the grain boundaries is especially pronounced for doped BaTiO$_3$ materials which are used as PTC resistors. Basically two types of dopants can be introduced into BaTiO$_3$: large ions of valence 3+ and higher, can be incorporated into Ba^{2+} positions, while the small ions of valence 5+ and higher, can be incorporated into the Ti^{4+} sublattice [9-11]. Usually, the extent of the solid solution of a dopant ion in a host structure depends on the site where the dopant ion is incorporated into the host structure, the compensation mechanism and the solid solubility limit [12]. For the rare-earth-ion incorporation into the BaTiO$_3$ lattice, the BaTiO$_3$ defect chemistry mainly depends on the lattice site where the ion is incorporated [13]. It has been shown that the three-valent ions incorporated at the Ba^{2+} -sites act as donors, which extra donor charge is compensated by ionized Ti vacancies ($V_{Ti}^{'''}$), the three-valent ions incorporated at the Ti^{4+} -sites act as acceptors which extra charge is compensated by ionized oxygen vacancies ($V_O^{\cdot\cdot}$), while the ions from the middle of the rare-earth series show amphoteric behavior and can occupy both cationic lattice sites in the BaTiO$_3$ structure [12]. It has been established that Ba/Ti ratio also influences the incorporation of rare earth dopants into the barium titanate lattice [14]. The investigations of the influence of Er on BaTiO$_3$ defect chemistry pointed out that for Ba/Ti>1 Er3+ enters on the titanium site, and for Ba/Ti < 1, it enters on the barium site. Moreover, erbium incorporation on barium or titanium sites introduces a charge mismatch with the lattice (an aliovalent dopant) that must be charge compensated to achieve overall neutrality [15]. As a result of rare-earth addition, the abnormal grain growth and the formation of deep and shallow traps at grain boundaries influenced by the presence of an acceptor-donor dopant can be observed.

Our investigations showed that the microstructure of the samples doped with Er$_2$O$_3$ or Yb$_2$O$_3$ exhibit similar microstructure characteristics with the existence of intergranular capacity. The samples sintered with Er$_2$O showed that the grains were irregularly polygonal shaped (Fig. 2a), although in Yb doped BaTiO$_3$ the grains are more spherical in shape (Fig.2.b).

For the lowest concentration, the size of the grains was large (up to 60 μm), but by increasing the dopant concentration the grain size decreased. As a result, for 0.1 wt% of dopant the average grain size was from 20 to 30 μm, and for the samples doped with 0.5 wt% of dopant grain size decreased to the value up to 10 μm as can be seen from the cumulative grain size distribution curves for doped BaTiO$_3$, given in Fig 3.

Spiral concentric grain growth has been noticed for the samples sintered with low concentration of Er$_2$O$_3$ or Yb$_2$O$_3$. For these samples the formation of the "glassy phase" indicated that the sintering was done in liquid phase.

EDS analysis has been shown that for the small concentration of Er and Yb, the uniform distribution has been noticed (Fig.4a), while the increase of dopant concentration led to the coprecipitation between grains (Fig.4b).

As a result the region between the grains can be represented by an electrical equivalent network consisting of three RC branches as it has been noticed in introduction.

All this allowed us to consider BaTiO$_3$-ceramics sample as a system with a huge number of mutually contacted grains which form clusters. For each of them, it is possible to establish the equivalent electrical model and, for defined set of input parameters, using symbolic analysis, obtain the frequency diagram. Realizing the totality of relations between clusters grains groups, their microelectrical schemes and corresponding frequency characteristics, from one side, and global equivalent electrical scheme and corresponding acquired frequency characteristics of BaTiO$_3$-ceramics samples, on the other side, we set a goal of coinciding experimental results with the summing effect of microelectric equivalent schemes.

According to the microstructures we have obtained for BaTiO$_3$ doped with Er$_2$O$_3$ and Yb$_2$O$_3$ it can be concluded that the global impedance of barium-titanate ceramics sample, which contains both resistor and capacity component, can be presented as a "sum" of many clusters of micro-resistors and micro-capacitors connected in tetrahedral lattice. According to our previous results, for the general model of the stereological configuration of BaTiO$_3$, consisting of a cluster of spheres or ellipsoids, density of the sample can be defined by mutual positions of the neighboring grains (Fig. 5).

Developed model gives the distribution of grain contacts through the sample volume. As it can be seen from the typical spatial situation of four grains cluster (Fig 6), the positions of the neighboring grains can be: 1. in touching contact; 2. slightly immerging one into another and 3. not touching each other.

This configuration logically leads to the tetrahedral scheme of mutual electrical influence of BaTiO$_3$ grains. The impedances are at the each edge of tetrahedron, as it is shown on Fig. 7. The vertices (in Fig. 7 displayed as small spheres) are stylized grains, while impedances contain resistance and capacity between two grains.

Our new approach includes fractal geometry in describing complexity of the spatial distribution of BaTiO$_3$ grains. The best fractal model is a sponge model, or and it is more correct to say, a kind of three –dimensional lacunary set (a set with voids). The structure of tetrahedral influence may be established in each spatial sense, which means that one has a tetrahedral lattice that fills the space.

This kind of fractals sometimes is identified with deterministic constructions like *Cantor set* (in P), *Sierpinski triangle* or *Sierpinski square* (in P^2), with *Sierpinski pyramid* or *Menger sponge* (in P^3), and so on. In our case of electric impedance property,

The *Sierpinski pyramid* (Fig. 8) might be a quite adequate paradigm for the first instance inquiry. The starting pyramid $\mathbf{T}_0$ and the first two iterations, shown in Fig. 7, give initial part of the orbit of so called Hutchinson operator W, that is $\mathbf{T}_1 = W(\mathbf{T}_0)$, and $\mathbf{T}_2 = W(\mathbf{T}_1) = W^2(\mathbf{T}_0)$. The limit case, $\mathbf{T} = W^\infty (\mathbf{T}_0)$ is an exact fractal set with Hausdorff dimension $D_H = 2$.

Following this model of Sierpinski pyramid, the induced model of impedances between clusters of ceramics grains is displayed in Fig 9. The task is to calculate equivalent impedance for the pyramid $\mathbf{T}_k$ in function of k.

Some of the impedances in this lattice are missing following the lacunary fractal pattern. According to [16], [17], the fractal dimension of the typical ceramics sample is estimated to be about 2.087. So in this case, that fractal dimension of *BaTiO$_3$* –ceramics is slightly below 3. The higher values of fractal dimensions can be obtained for more dence *BaTiO$_3$* –ceramics. The dimension close to 3 means that not many of the impedances lack in the tetrahedral lattice. On the other hand, dimension closer to 2.5 alerts high presence of pores which means that many impedances are to be swept in the model of tetrahedral impedance lattice. What is sure is that BaTiO$_3$ sintered ceramics can be considered as a lacunar fractal rather than a percolation fractal or succolar fractal [18].

In order to calculate equivalent impedance for a wide frequency range, the equivalent electrical circuit for a ceramic material can be introduced as an impedance containing two capacitance C and C$_p$, an inductance L and a resistance R.

The dominant electrical parameter of our model is the capacitance C. The connection between C and geometrical and/or structural properties can be established by an assumption that the contact region can be viewed as a planar micro-capacitance. Fig. 10. shows a geometric model of two spheres in contact, where r$_c$ is the radius of the spherical particles, and x is the neck radius.

Since the spherical shape and the circular neck profile are assumed, the contact surface is the circle of the area πx^2. Because of center-to center approach, the dielectric thickness 2h is a function of d (the distance between centers of particles) according to the relatio 2h=2r$_c$-d. Therefore, the capacitance can be written as:

$$C = \varepsilon_o \varepsilon_r \alpha \frac{\pi x^2}{2r_c - d} \quad , \quad (d = 2\sqrt{r_c^2 - x^2}) \tag{1}$$

where ε_o and ε_r are the diuletric constants of vacuum and the ceramic material, respectively and α is a correction factor obtained by a constructive approach to the fractal structure. Taking into account that by the fractal theory α can be presented as:

$$\alpha = D - D_T \tag{2}$$

where D$\approx$2.08744 is the fractal (Hausdorff) dimension of intergrain contact surface and D$_T$=2 is topological dimension of the surface, it can be concluded that for BaTiO$_3$ doped ceramics contact surfaces are of low-irregularity which is characterized by the small diference D - D$_T$ $\approx$0.08744.

Taking this into account, calculations of microcapacitance generated in grains contacts of BaTiO$_3$ doped with Er$_2$O$_3$ and Yb$_2$O$_3$ have been performed.

In order to calculate equivalent impedance of the sample, conductance as another dominant electrical parameter shuold be taken into account. It is a parasitic parameter that is given in terms of capacitance with tanδ as a measure of losses, i.e., 1/R=G=ωC·tanδ. The intergranular impedance model also contains two additional parameters; inductance L, and capacitance C$_p$. Their nature cannot be correlated with geometrical parameters of grains in general way.

In order to determine an algebaric equation describing equivalent intergranular impedance in terms of circuit parameters following equation can be used:

$$Z(s) = \frac{1 + CR \cdot s + CL \cdot s^2}{(Cp + C) \cdot s + CpCR \cdot s^2 + CpCL \cdot s^3} \tag{3}$$

Based on proposed equivalent circuit and the theory of impedance analysis for the model of three aggregate spheres the equivalent impedance can be defined by:

$$Z_e = \frac{Z_{12} \cdot (Z_{13} + Z_{23})}{Z_{12} + Z_{13} + Z_{23}} \tag{4}$$

where Z_{12}, Z_{13}, Z_{23} are the intergranular impedances between two adjacent particles. Then, this model can be inserted for any contact region inside the multi-particle model system during its microstructure development. Thus, electrical properties are determined in general by a series combination of such impedances.

CONCLUSION

Understanding the electrical properties of barium-titanate materials is important for modern devices applications and present a challenge for their simulation. In this study, the model of intergranular impedance is established using the equivalent electrical scheme characterized by corresponding frequency characteristic. According to the microstructures we have obtained for BaTiO₃ doped with Er₂O₃ and Yb₂O₃ the global impedance of barium-titanate ceramics sample, which contains both resistor and capacity component. Resistor and capacity component has been presented as a "sum" of many clusters of micro-resistors and micro-capacitors connected in tetrahedral lattice. The positions of neighboring grains for the four grains cluster have been defined and according to them the tetrahedral scheme of mutual electrical influence of BaTiO₃ grains has been established. Fractal geometry has been used to describe complexity of the spatial distribution of BaTiO₃ grains. The model of impedances between clusters of ceramics grains has been presented and calculations of microcapacitance generated in grains contacts of BaTiO₃ doped with Er₂O₃ and Yb₂O₃ have been performed. By the control of shapes and numbers of contact surfaces on the level of the entire BaTiO₃–ceramic sample, the control over structural properties of this ceramics can be done, with the aim of correlation between material electronic properties and corresponding microstructure.

Acknowledgements: This research is a part of the project "Investigation of the relation in triad: synthesis-structure-properties for functional materials" (No.142011G). The authors gratefully acknowledge the financial support of Serbian Ministry for Science for this work.

REFERENCES

[1] M.M.Vijatovic, J.D.Bobic, B.D.Stojanovic History and Challenges of Barium Titanate I Sci.Sint Vol.40 2 (2008) 155-167

[2] C.Pithan, D.Hennings, R. Waser Progress in the Synthesis of Nanocrystalline BaTiO₃ Powders for MLCC International Journal of Applied Ceramic Technology 2 (1), (2005), 1–14.

[3] V.V. Mitić, I. Mitrović, D. Mančić, "The Effect of CaZrO₃ Additive on Properties of BaTiO₃-Ceramics", Sci. Sint., Vol. 32 (3), pp. 141-147, 2000

[4] P.W.Rehrig, S.Park, S.Trolier-McKinstry, G.L.Messing, B.Jones, T.Shrout Piezoelectric properties of zirconium-doped barium titanate single crystals grown by templated grain growth J. Appl. Phys. Vol 86 3, (1999) 1657-1661

[5] S. Wang, G.O. Dayton Dielectric Properties of Fine-Grained Barium Titanate Based X7R Materials J. Am. Ceram. Soc. 82 (10), (1999), 2677–2682.

[6] V.P.Pavlovic, M.V.Nikolic, V.B.Pavlovic, N. Labus, Lj. Zivkovi}, B.D.Stojanovic, Correlation between densification rate and microstructure evolution of mechanically activated BaTiO$_3$, Ferroelectrics 319 (2005) 75-85

[7] D. Lu, X. Sun, M. Toda Electron Spin Resonance Investigations and Compensation Mechanism of Europium-Doped Barium Titanate Ceramics Japanese Journal of Applied Physics Vol. 45, No. 11, 2006, pp. 8782-8788

[9] M. Vollman, R. Waser Grain Boundary Defect Chemistry of Acceptor-Doped Titanates: Space Charge Layer Width J.Am.Ceram.Soc. 77[1] 235-43 (1994)

[10] V.V. Mitić, I. Mitrović, "The Influelnce of Nb$_2$O$_5$ on BaTiO$_3$.Ceramics Dielectric Properties", Journal of the European Ceramic Society, Vol. 21 (15), pp. 2693-2696, 2001.

[11] H.M.Chan, M.P.Hamer, D.M.Smyth, Compensating defects in highly donor-doped BaTiO 3. J.Am. Ceram. Soc., 69(6) (1986) 507-10.

[12] P.W.Rehrig, S.Park, S.Trolier-McKinstry, G.L.Messing, B.Jones, T.Shrout Piezoelectric properties of zirconium-doped barium titanate single crystals grown by templated grain growth J. Appl. Phys. Vol 86 3, (1999) 1657-1661.

[13] D.Makovec, Z.Samardzija M.Drofenik Solid Solubility of Holmium, Ytrium and Dysprosium in BaTiO$_3$ J.Am.Ceram.Soc. 87 [7] 1324-1329 (2004).

[14] D. Lu, X. Sun, M. Toda Electron Spin Resonance Investigations and Compensation Mechanism of Europium-Doped Barium Titanate Ceramics Japanese Journal of Applied Physics Vol. 45, No. 11, 2006, pp. 8782-8788.

[15] M. T. Buscaglia, V. Buscaglia, P. Ghigna, M. Viviani, G. Spinolo, A. Testino and P. Nanni, Amphoteric behaviour of Er^{3+} dopants in BaTiO$_3$: an Er–L$_{III}$ edge EXAFS assessment *Phys. Chem. Chem. Phys.*, 2004, **6**, 3710

[16] John D. Bak and John C. Wright Site-Selective Spectroscopy of the Solid-State Defect Chemistry in Erbium-Doped Barium Titanate *J. Phys. Chem. B* 2005, *109*, 18391-18399

[17] V.V.Mitić, Lj. M. Kocić, M. Miljković and I. Petković, Fractals and BaTiO$_3$ – Ceramics microstructure analysis, Mikrochim. Acta [Suppl.] **15**, (1998), 365-369

[18] V.Mitic Lj.Kocic, I.Mitrovic, M.M.Ristic Models of BaTiO$_3$ Ceramics Grains Contact Surfaces The 4th IUMRS International Conference in Asia OVTA Makuhari, Chiba, Japan 1997

[20] B. Mandelbrot, *The Fractal Geometry of Nature*, W. H. Freeman and Co., New York, 1983.

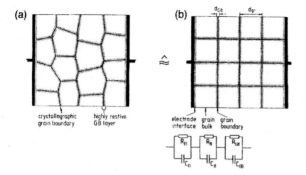

Fig. 1. Sketched cross section of a ceramic sample with electrodes applied (a), corresponding simplified brick wall model (b), and the equivalent electrical network

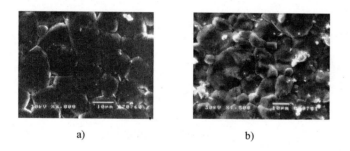

Fig.2. SEM micrograph of doped BaTiO₃ sintered at 1320°C a) 0.5Er-BT and   b) 0.5Yb-BT.

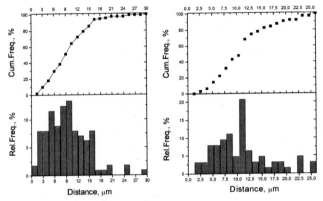

Fig. 3. Cumulative grain size distribution curves for doped BaTiO$_3$ with a) 0.5 wt% Er$_2$O$_3$, and b) 0.5 wt% Yb$_2$O$_3$.

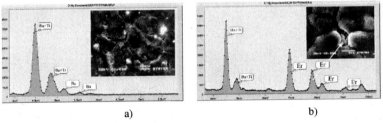

a)   b)

Fig.4. SEM/EDS spectra of doped BaTiO$_3$ sintered at 1320°C a) 0.1Er-BT and   b) 0.5Er-BT

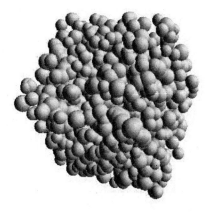

Fig. 5. The simulation of the 1000 grains of BaTiO$_3$ obtained from the general model of the stereological configuration.

Fig. 6. The positions of neighboring grains for the four grains cluster.

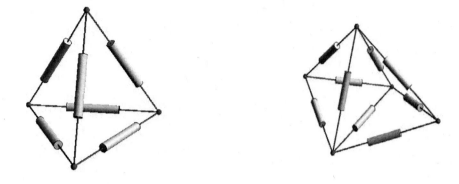

Fig. 7. The impedance model of tetrahedron.

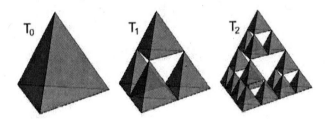

Fig. 8. The Sierpinski pyramid.

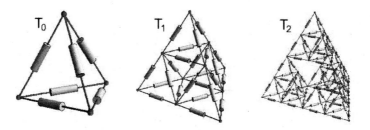

Fig. 9. The model of impedances between clusters of ceramics grains.

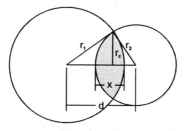

Fig .10 Two-sphere contact model presented in plane section.

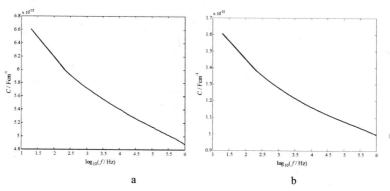

a   b

Fig. 11. Microcapacitance vs. frequency for BaTiO$_3$ doped with
a) 0.5 wt% Er$_2$O$_3$ b) 0.5 wt% Yb$_2$O$_3$.

Thermal to
Electric Conversion

Ca-DOPING AND THERMOELECTRIC PROPERTIES OF Ca_xCoO_2 EPITAXIAL FILMS

Kenji Sugiura,[1] Hiromichi Ohta,[1] Kenji Nomura,[2] Hideo Hosono,[2,3] and Kunihito Koumoto [1,4]

[1]*Graduate School of Engineering, Nagoya University, Furo-cho, Chikusa, Nagoya 464-8603, Japan*
[2]*ERATO-SORST, Japan Science and Technology Agency (JST), in Frontier Collaborative Research Center (FCRC), Tokyo Institute of Technology, Mail box S2-13, 4259 Nagatsuta, Midori-ku, Yokohama 226-8503, Japan*
[3]*Materals and Structures Laboratory, Mail Box R3-1, and FCRC, Mail box S2-13, Tokyo Institute of Technology, 4259 Nagatsuta, Midori-ku, Yokohama 226-8503, Japan*
[4]*CREST, JST, Tokyo 332-0012, Japan*

ABSTRACT

We have measured the thermoelectric properties of layered Ca_xCoO_2 epitaxial films when the Ca-content was varied from $x = 0.33$ to 0.45. The $Ca_{0.33}CoO_2$ film, which was fabricated by ion-exchange of $Na_{0.7}CoO_2$ film, successfully incorporated the additional Ca-ions in-between the CoO_2 layers by the thermal annealing with $Ca(OH)_2$ powder. The thermal annealing caused the Ca-content to increase up to $x = 0.45$ after the transformation of Ca-arrangements. The electrical conductivity of the film changed from metallic behavior to semiconducting behavior by the transformation of Ca-arrangement, and became more than two orders of magnitude lower with increasing x. The Seebeck coefficient of the film gradually increased with increasing x, and reached to ~ 210 $\mu V \cdot K^{-1}$ at $x = 0.45$.

I. INTRODUCTION

A series of layered cobaltates, including Na_xCoO_2, Sr_xCoO_2, and $Ca_3Co_4O_9$ [1-6], have attracted attention as a candidate for p-type thermoelectric (TE) power generation material. Generally, TE performance is characterized by means of figure of merit, ZT $(= S^2 \cdot \sigma \cdot T \cdot \kappa^{-1}$, where S, σ, T and κ are Seebeck coefficient, electrical conductivity, absolute temperature, and thermal conductivity, respectively), and the cobaltates can simultaneously exhibit a fairly large S (~ 100 $\mu V \cdot K^{-1}$ at 300 K) and metallic σ possibly due to the strong electron correlation effect [1,7-9]. Since these factors in TE performance can be strongly influenced by carrier concentration [10], clarification of the relationships between TE properties and carrier concentration is essential for exploration of a high efficiency TE cobaltate.

A layered cobaltate generally consists of alternating stacks of a CoO_2 layer and an alkali/alkaline-earth metal layer along the [0001]-direction as shown in Fig. 1(a). The number of positive hole carriers increase/decrease by subtracting/adding alkali or alkaline-earth cations because oxidation/reduction of cobalt-ions between Co^{3+} and Co^{4+} occurs simultaneously. In Na_xCoO_2, the Na-content, which is about $x \sim 0.7$ in as-grown crystal, is varied from 0.3 to ~ 1.0 by chemical processes [2,11,12]. Recently, Lee et al. [2] reported the relationship between x and S in Na$_x$CoO$_2$ and observed the large enhancement of S in high Na-doping region of Na$_x$CoO$_2$ ($x > 0.75$). Although the Na$^+$-incorporated Na$_x$CoO$_2$ is a good example to clarify the correlation between TE properties and cation-contents in the layered cobaltates system, it is very difficult to handle Na$^+$-including cobaltates, especially highly Na-doped one, because it may be unstable even at room temperature [2,4].

In the present study, we focused our attention on Ca$_x$CoO$_2$ [13] to investigate the cation-content dependence of TE properties. This is because the divalent metal cation Ca^{2+} makes the Co-valence varied largely with lower cation-contents. Moreover, it has been observed that Ca$_x$CoO$_2$ has some vacancies of Ca-ions in a Ca-ordered structure [14]. This fact makes us expect the possibility that further Ca-ions are intercalated in Ca$_x$CoO$_2$. In addition, high-quality epitaxial films of Ca$_x$CoO$_2$ are successfully fabricated via ion-exchange of Na$_x$CoO$_2$ epitaxial films [15,16]. The high-quality epitaxial film is useful to clarify some of relevant TE properties. Herein we report the Ca-doping and TE properties of Ca$_x$CoO$_2$ epitaxial films. The prepared Ca$_x$CoO$_2$ film had the Ca-content of $x = 0.33$, and then further Ca-ions were successfully intercalated in the film up to $x = 0.45$ by the thermal annealing with Ca(OH)$_2$ powder.

II. EXPERIMENTAL METHODS

Ca$_x$CoO$_2$ ($x = 0.33$) epitaxial films were fabricated by an ion-exchange treatment of Na$_{0.7}$CoO$_2$ epitaxial films, which were grown on the (0001)-face of α-Al$_2$O$_3$ substrates by reactive solid-phase epitaxy (R-SPE [17,18]). The details of the R-SPE and subsequent ion-exchange treatment are described elsewhere [15,16,19,20]. The prepared Ca$_{0.33}$CoO$_2$ epitaxial film has the following epitaxial relationship with the α-Al$_2$O$_3$ substrate: (0001)[11$\bar{2}$0] Ca$_{0.33}$CoO$_2$ $\|$ (0001)[1$\bar{1}$00] α-Al$_2$O$_3$ [15,16]. In order to intercalate further Ca-ions into the Ca$_{0.33}$CoO$_2$ film, the film was annealed with Ca(OH)$_2$ powder. The appropriate amount of Ca(OH)$_2$ powder was placed directly on the Ca$_{0.33}$CoO$_2$ film, and the film was subsequently heated below 700°C for 1 h in air. After heating, the film was washed with distilled water to remove the residual powder on the film.

The chemical composition of the resultant film was confirmed by X-ray fluorescence (XRF; ZSX 100e, Rigaku Co.) analysis. The crystallographic orientation, lattice parameters, and film thickness were evaluated by high-resolution x-ray diffraction (HR-XRD; ATX-G, Rigaku-Co.) measurements. σ and S were measured by the dc four-point probe method and the conventional steady state method, respectively. Sputtered Au electrodes were used for ohmic contacts.

III. RESULTS AND DISCUSSIONS

Figure 2 shows (a) out-of-plane and (b) in-plane HR-XRD patterns of the Ca$_{0.33}$CoO$_2$ film after annealing with Ca(OH)$_2$ powder at each temperature (annealing temperature: $T_a = 25$℃ (as-prepared), 300℃, and 700℃). Only the intense 000l peaks from Ca$_x$CoO$_2$ were observed with the 0006 peak from α-Al$_2$O$_3$ in the out-of-plane patterns [Fig. 2(a)]. Throughout the annealing treatment, a significant change was not observed in their patterns, indicating that the layered structure composed of CoO$_2$ lattices is kept without depositing other crystalline phases. It should be noted that the annealing of the film above 800℃ brought about the transformation from Ca$_x$CoO$_2$ to Ca$_3$Co$_4$O$_9$ [15]. In the in-plane patterns of the film [Fig. 2(b)], the superstructure peaks from the Ca-ordered structures are observed with $11\bar{2}0$ peak from the Ca$_x$CoO$_2$ film and the $3\bar{3}00$ peak from α-Al$_2$O$_3$ [13,16]. Two superstructure peaks were observed at the 1/3 and 2/3 × q_x of $11\bar{2}0$ peak in the pattern of the as-prepared film (T_a = 25℃), but the two superstructure peaks were replaced by a single peak at the 1/2 × q_x of the $11\bar{2}0$ peak above $T_a \geq 300°C$. This change of the pattern indicates the transformation of Ca-arrangements. The pattern at T_a = 25℃ is attributed to the $\sqrt{3}a \times \sqrt{3}a$ hexagonal Ca-ordered structure whereas the patterns in $T_a \geq 300°C$ are attributed to the $2a \times \sqrt{3}a$ orthorhombic one. The details of the superstructure peaks in the in-plane XRD are described elsewhere (ref. 16). These Ca-ordered structures, shown in Fig. 1(b), had been confirmed by high angle annular dark-field scanning transmission electron microscopy (HAADF-STEM) observations [14]. These XRD results indicate that the annealing treatment of the film above 300℃ caused the Ca-arrangement of Ca$_x$CoO$_2$ to be varied from the $\sqrt{3}a \times \sqrt{3}a$ hexagonal type to the $2a \times \sqrt{3}a$ orthorhombic type.

Figure 3 shows (a) the lattice constants, c, a, and (b) the Ca-content, x of the Ca$_x$CoO$_2$ film as a function of T_a. The c- and a-values were obtained from the out-of-plane and in-plane XRD peaks, respectively. Although the Ca-arrangement was varied near T_a = 300℃, the Ca-content of the film did not change (x = 0.33) and the lattice constants were nearly constant below T_a = 400℃. Because the $2a \times \sqrt{3}a$ orthorhombic structure should have a cation site of x = 0.5 [13], this experimental result suggests that the Ca-content is insufficient to occupy the Ca-sites in the orthorhombic structure. In fact, the existence of Ca-vacancies has been confirmed by HAADF-STEM observation [14]. The Ca-site occupancy was estimated as ~66% from the Ca-content (= 0.33) / the Ca-site (= 0.5). Hence, one can expect that the orthorhombic Ca-arrangement with some Ca-vacancies makes it possible to intercalate further Ca-ions in the film. Actually, above T_a = 500℃, the Ca-content was found to increase in Fig. 3(b). That gradually increased with increasing T_a and reached to x = 0.45 at T_a = 700℃. With increasing the Ca-content of the film, the lattice constants were also varied slightly [Fig. 3(a)]. Then, during the increase of Ca-content, decomposition of the film or deposition of other crystalline phases were not observed as already seen in Fig. 2. These results indicate that the Ca-ions were successfully intercalated in the film by the annealing with Ca(OH)$_2$ powder above 500℃, and that the Ca-content was varied from x = 0.33 to 0.45.

Figure 4(a) shows the temperature dependence of σ for the Ca$_x$CoO$_2$ thin films (90-nm-thick) with different x-values, which is represented by the Arrhenius plot. The Ca$_{0.33}$CoO$_2$ film with the hexagonal Ca-arrangement (T_a = 25℃) displayed metallic conductivity and the σ-value was as high as 1.4 ×10^3 S·cm^{-1} at 300 K. However, with the transformation of Ca-arrangements to the orthorhombic type (T_a = 400℃), σ changed to semiconducting behavior even though the same Ca-content (x = 0.33). The σ-value became about one order of magnitude lower (4.8 ×10^1 S·cm^{-1}) at 300 K. The details of the metal-insulator transition with the transformation of Ca-arrangements in the Ca$_{0.33}$CoO$_2$ film are described in ref. 16. Then, σ of the film was further decreased with increasing x as seen in Fig. 4(a). The σ-value became about two orders of magnitude lower as the Ca-content changed from $x = 0.33$ to 0.45, and reached to 1.1 ×10^{-1} S·cm^{-1} (300 K) at the $x = 0.45$. Figure 4(b) shows the temperature dependence of S for the Ca$_x$CoO$_2$ films with different x-values. The Ca$_{0.33}$CoO$_2$ film with the hexagonal Ca-arrangement ($T_a = 25$℃) exhibited metallic temperature-dependence of S and the S-value was 90 μV·K^{-1} at 300 K. The S-values merely exhibited a slight increase with the transformation of Ca-arrangements from the hexagonal type to orthorhombic type possibly because of disorder effect [16], but those increased largely with increasing x. The S-value of Ca$_{0.45}$CoO$_2$ at 300 K was as high as ~210 μV·K^{-1}, which is about two times larger than that of Ca$_{0.33}$CoO$_2$.

The observed x-dependence of σ and S suggests that Ca-doping systematically vary the fraction of Co^{3+}/Co^{4+} mixed-valence state (~ hole concentration). Figure 5(a) shows the x-dependence of S at 300 K for Ca$_x$CoO$_2$, showing that the S-value gradually increases with increasing x. According to Koshibae et $al.$ [7], in the cobaltates system with Co^{3+}/Co^{4+} mixed-valence state, S near room temperature depends on the fraction of Co^{3+}/Co^{4+} mixed-valence as expressed by the following equation:

$$S = -\frac{k_B}{e} \ln\left(\frac{g_3}{g_4} \frac{y}{1-y}\right) \qquad (1)$$

where k_B is the Boltzmann constant, e is the elementary charge, g_3 and g_4 are the spin-orbital degeneracies of Co^{3+} and Co^{4+}, respectively, and y is the fraction of Co^{4+} ions ($y = 1 - 2x$). Since there are two potential spin-orbital degeneracies: $g_3 = 1$, $g_4 = 2$, corresponding to the single band hopping model, and $g_3 = 1$, $g_4 = 6$, which corresponds to the correlated hopping model of low-spin Co^{3+}/Co^{4+} mixed-valence state [21,22], we have plotted both x-S curves calculated from eq. (1) in Fig. 5(a). If we assume single band hopping ($g_3 = 1$, $g_4 = 2$), the trend of x-S relation calculated from eq. (1) agrees well with that observed in Ca$_x$CoO$_2$. This coincidence confirms the successful Ca-doping by the annealing of the film with Ca(OH)$_2$ powder. The fact that the single band hopping model is better with the experimentally observed S-values is likely due to the rhombohedral distortion of the CoO$_6$-octahedra in the CoO$_2$ layer, which makes the Co t_{2g} bands separated into a_{1g} and e'_g bands. In fact, according to the photoemission study of Na$_x$CoO$_2$ [23,24], the density of state near E_F predominantly consists of a_{1g} bands and the e'_g bands sink below E_F (several hundred meV), though the amount of splitting depends on the extent of the distortion.

The above results present the TE properties of Ca$_x$CoO$_2$ epitaxial films when the Ca-content was successfully varied. Finally, we'd compare the σ- and S-values of the present Ca$_x$CoO$_2$ films with those reported Na$_z$CoO$_2$ [2,12], where z is corresponds to $2x$ due to standardization of the Co valence between Ca$_x$CoO$_2$ and Na$_z$CoO$_2$. The x-dependence of S at 300 K for Ca$_x$CoO$_2$ is almost similar with that for Na$_z$CoO$_2$ [2,12] as shown in Fig. 5(a). This suggests that S is generally dominated by the Co-valence state in this system. On the other hand, x-dependence of σ shows a large difference between Ca$_x$CoO$_2$ and Na$_z$CoO$_2$. Figure 5(b) shows the comparison of the σ-value at 300 K between Ca$_x$CoO$_2$ and Na$_z$CoO$_2$. As observed in Fig. 4(a), Ca$_x$CoO$_2$ displayed a significant change of σ due to the transformation of Ca-arrangements at $x = 0.33$, and then the σ-value became as low as ~10^{-1} S·cm^{-1} with increasing x. Na$_z$CoO$_2$ also shows the decrease of σ with increasing Na-content, but magnitude of the change is smaller and moderately large σ (~10^1 S·cm^{-1}) is kept even in the vicinity of $z = 1$, where almost all Co-ions are Co^{3+} and hole concentration seems to be rather low. Furthermore, Na$_z$CoO$_2$ always shows metallic conductivity [2,12] except for $x = 0.5$ [12]. These differences of σ suggest that the conductivity in this system is sensitive to sorts, arrangements, or distribution of cations between CoO$_2$ layers.

IV. CONCLUSIONS

We successfully intercalated the additional Ca-ions in the Ca$_{0.33}$CoO$_2$ epitaxial film by the annealing with Ca(OH)$_2$ powder. The thermal annealing caused the Ca-arrangement of Ca$_{0.33}$CoO$_2$ to be varied from a $\sqrt{3}a \times \sqrt{3}a$ hexagonal type to a $2a \times \sqrt{3}a$ orthorhombic type with the identical Ca-content, and to incorporate the additional Ca-ions up to $x = 0.45$. σ of the film changed from metallic behavior to semiconducting behavior by the transformation of Ca-arrangement, and became more than two orders of magnitude lower with increasing x. S of the film gradually increased with increasing x, and reached to ~210 μV·K^{-1} at $x = 0.45$. To develop a high efficiency TE cobaltate, it may be important to control the doping concentration with holding metallic conductivity.

REFERENCES

[1] I. Terasaki, Y. Sasago, and K. Uchinokura, *Phys. Rev. B* **56**, R12685 (1997).

[2] M. Lee, L. Viciu, L. Li, Y. Wang, M. L. Foo, S. Watauchi, R. A. Pascal Jr, R. J. Cava, and N. P. Ong, *Nat. Mater.* **5**, 537 (2006).

[3] R. Ishikawa, Y. Ono, Y. Miyazaki, and T. Kajitani, *Jpn. J. Appl. Phys., Part 2* **41**, L337 (2002).

[4] K. Sugiura, H. Ohta, K. Nomura, M. Hirano, H. Hosono, and K. Koumoto, *Appl. Phys. Lett.* **88**, 082109 (2006).

[5] A. C. Masset, C. Michel, A. Maignan, M. Hervieu, O. Toulemonde, F. Studer, and B. Raveau, *Phys. Rev. B* **62**, 166 (2000).

[6] M. Shikano and R. Funahashi, *Appl. Phys. Lett.* **82**, 1851 (2003).

[7] W. Koshibae, K. Tsutsui, and S. Maekawa, *Phys. Rev. B* **62**, 6869 (2000).

[8] D. J. Singh, *Phys. Rev. B* **61**, 13397 (2000).

[9] Y. Ishida, H. Ohta, A. Fujimori, and H. Hosono, *J. Phys. Soc. Jpn.* **76**, 103709 (2007).

[10] G. Mahan, B. Sales, and J. Sharp, *Phys. Today* **50**, 42 (1997).

[11] S. Kikkawa, S. Miyazaki, and M. Koizumi, *J. Solid State Chem.* **62**, 35 (1986).

[12] M. L. Foo, Y. Wang, S. Watauchi, H. W. Zandbergen, T. He, R. J. Cava, and N. P. Ong, *Phys. Rev. Lett.* **92**, 247001 (2004).

[13] H. X. Yang, Y. G. Shi, X. Liu, R. J. Xiao, H. F. Tian, and J. Q. Li, *Phys. Rev. B* **73**, 014109 (2006).

[14] R. Huang, T. Mizoguchi, K. Sugiura, H. Ohta, K. Koumoto, T. Hirayama, and Y. Ikuhara, *Appl. Phys. Lett.* **93**, 181907 (2008).

[15] K. Sugiura, H. Ohta, K. Nomura, M. Hirano, H. Hosono, and K. Koumoto, *Appl. Phys. Lett.* **89**, 032111 (2006).

[16] K. Sugiura, H. Ohta, Y. Ishida, R. Huang, T. Saito, Y. Ikuhara, K. Nomura, H. Hosono, and K. Koumoto, *Submitted.*

[17] H. Ohta, K. Nomura, M. Orita, M. Hirano, K. Ueda, T. Suzuki, Y. Ikuhara, and H. Hosono, *Adv. Funct. Mater.* **13**, 139 (2003).

[18] K. Nomura, H. Ohta, K. Ueda, T. Kamiya, M. Hirano, and H. Hosono, *Science* **300**, 1269 (2003).

[19] H. Ohta, S.-W. Kim. S. Ohta, K. Koumoto, M. Hirano, and H. Hosono, *Cryst. Growth Des.* **5**, 25 (2005).

[20] H. Ohta, A. Mizutani, K. Sugiura, M. Hirano, H. Hosono, and K. Koumoto, *Adv. Mater.* **18**, 1649 (2006).

[21] P. M. Chaikin and G. Beni, *Phys. Rev. B* **13**, 647 (1976).

[22] M. Pollet, J. P. Doumerc, E. Guilmeau, D. Grebille, J. F. Fagnard, and R. Cloots, *J. Appl. Phys.* **101**, 083708 (2007).

[23] M. Z. Hasan, Y. D. Chuang, D. Qian, Y. W. Li, Y. Kong, A. Kuprin, A. V. Fedorov, R. Kimmerling, E. Rotenberg, K. Rossnagel, Z. Hussain, H. Koh, N. S. Rogado, M. L. Foo, and R. J. Cava, *Phys. Rev. Lett.* **92**, 246402 (2004).

[24] D. Qian, L. Wray, D. Hsieh, L. Viciu, R. J. Cava, J. L. Luo, D. Wu, N. L. Wang, and M. Z. Hasan, *Phys. Rev. Lett.* **97**, 186405 (2006).

(a)

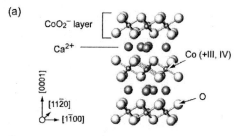

(b)

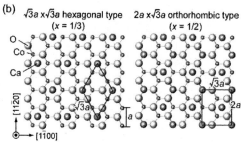

FIGURE 1. (a) Crystal structure of Ca_xCoO_2. (b) Two types of Ca-order structure observed in Ca_xCoO_2: $\sqrt{3}a \times \sqrt{3}a$ hexagonal type (left) and $2a \times \sqrt{3}a$ orthorhombic type (right). Note that the hexagonal order structure has a cation site of $x = 0.33$, whereas the orthorhombic one has a cation site of $x = 0.5$.

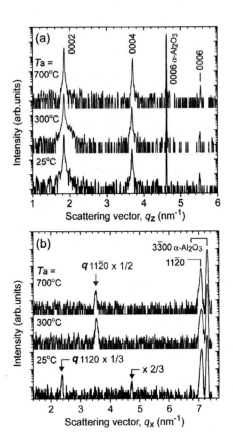

FIGURE 2. (a) Out-of-plane and (b) in-plane XRD patterns of Ca_xCoO_2 epitaxial films after annealing with $Ca(OH)_2$ powder at each temperature (T_a = 25°C (as-prepared), 300°C, and 700°C) in air. Arrows indicate the superstructure peaks from the Ca-ordered structures.

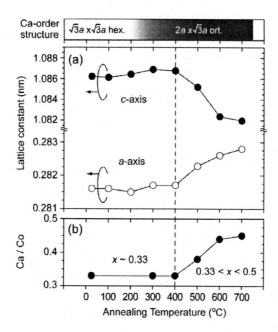

FIGURE 3. Lattice constants, a, c (a) and Ca-content, x (b) of Ca$_x$CoO$_2$ films after annealing with Ca(OH)$_2$ powder at each temperature below 700°C. Ca-order structures, judged from the superstructure pattern in the in-plane XRD, are also shown on the top.

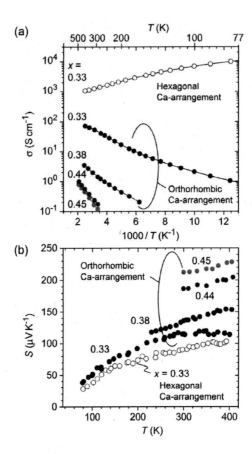

FIG. 4. (a) Log σ - 1000/T plot and (b) S - T curve for Ca$_x$CoO$_2$ films with different Ca-contents.

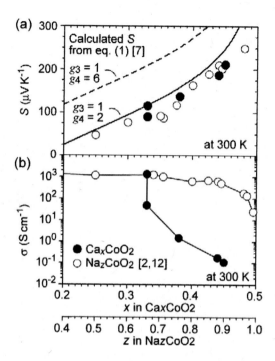

FIGURE 5. The cation-content dependence of (a) S- and (b) σ-values at 300 K for Ca$_x$CoO$_2$ films. Reported values for Na$_z$CoO$_2$ ($z = 2x$)[2,12] are also shown for comparison. The lines in (a) indicate the calculated S-values from eq. (1) (solid line: the single band model, dashed line: the correlated hopping model)[7].

THERMOELECTRIC GENERATORS

David Michael Rowe OBE DSc PhD.

Honorary Research Professor at Cardiff School of Engineering,
Queen's Buildings , Newport Road, Cardiff CF24 1XF,Wales,UK.

ABSTRACT

An introduction to thermoelectric generation is followed by a brief outline of the basic principles, which govern the thermoelectric conversion of heat flow into electricity.   Components of a thermoelectric generator are described together with their role.

Examples of generators are presented with power outputs ranging from micro-milliwatts to kilowatts and include an early Russian examples from the 1950's to   NASA's multihundred watt systems that provide on-board power to deep space missions.   Improvements   in generator performance due to advances in material figure- of- merit, thermoelement design   and module configurations are reviewed.

Current activity in thermoelectric generation is focused on waste heat recovery. Major sources of low and high temperature waste heat are identified. Efforts to develop vehicular exhaust waste heat recovery technology are reviewed and the potential of thermoelectric generation in combating global warming and reducing fuel consumption assessed.

The first crucial step in   transferring   the improvements in thermoelectric figure- of- merit obtained in nanostructured thin films to bulk material has been achieved This development in material research holds out the best opportunity to date for obtaining substantial improvements in bulk material performance. However, other factors such as material cost, weight, availably and toxicity need to be addressed before achieving   globalisation of thermoelectric generation and over a wide spectrum of applications.

INTRODUCTION

A thermoelectric generator (TEG) is a solid-state heat engine with the electron gas serving as the working fluid and converts a flow of heat into electricity. It has no moving components, is silent, totally scalable and extremely reliable. In the early 1960's a requirement for autonomous long–life sources of electrical power arose from the exploration of space, advances in medical physics, deployment of marine and terrestrial surveillance systems and the exploitation of the earth's resources in increasingly hostile and inaccessible locations[1]. Thermoelectric devices employing radioactive isotopes as a heat source (Radioisotope Powered Thermoelectric Generators, (RTGs) provided the required electrical power. Total reliability of this technology has been demonstrated in applications such as the Voyager space crafts. However, employing radioisotopes as sources of heat has remained restricted to specialised applications where the thermoelectric generator's desirable properties listed above outweighed its relatively low conversion efficiency (typically 5%). The fivefold increase in the price of crude oil in 1974, accompanied by   an increased awareness of environmental problems associated with global warming, resulted in an upsurge of scientific activity to identify and develop environmentally friendly sources of electrical power. Thermoelectric generation   in applications, which employ waste heat as a heat source, is a totally green technology and. over the past ten years or so effort has focused on developing thermoelectric generating systems which can harvest waste heat from the human body, computer chips,   industrial utilities and vehicle engines.

THERMOELECTRIC GENERATION, EFFICIENCY AND FIGURE-OF-MERIT.

A thermoelectric converter like all heat engines obeys the laws of thermodynamics. When operated in its generating mode   it will deliver electrical power to an external load provided it is supplied with heat energy.

Consider the converter operating as an ideal generator in which there are no heat losses - the efficiency is defined as the ratio of the electrical power delivered to the load to the heat absorbed at the hot junction. Expressions for the important parameters in thermoelectric generation can readily be derived by considering the simplest generator consisting of a single thermocouple with legs or thermo elements fabricated from n- and p-type semiconductors as shown in Figure 1 [2,3]

The efficiency of the generator is given by:

$$\phi = \frac{energy \, \sup plied \, to \, the \, load}{heat \, energy \, absorbed \, at \, hot \, junction}$$

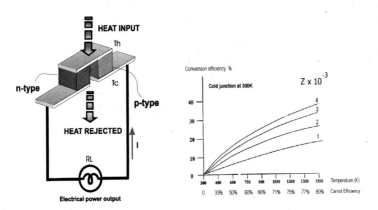

Figure 1. Single couple module   Figure 2. Efficiency as a function of Z and T.

The efficiency can be expressed as a function of the temperature over which the device is operated and a so-called 'goodness factor' or thermo-electric figure-of-merit (Z) of the thermocouple material.

$$Z = \frac{\alpha^2 \sigma}{\lambda}$$

Where $\alpha^2 \sigma$ is referred to as the electrical power factor, with α, the Seebeck coefficient, σ the electrical conductivity and λ is the total thermal conductivity. The figure-of-merit is often expressed in its dimensionless form ZT where T is absolute temperature.

In figure 3 is displayed the figure of merit as a function of temperature. The ZT maxima of all materials fall close to unity. A material with a Z value of 1×10^{-3} K^{-1} when operated with a hot side of 1000C would convert heat into electricity with efficiency in excess of 15%.

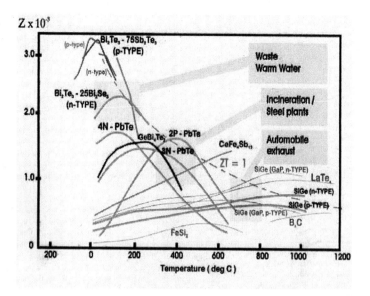

Figure 3. Z of established materials as a function of temperature

ESTABLISHED THERMOLECIC MATERIALS [4]

Established thermoelectric materials conveniently fall into three groupings with each dependent upon the temperature range of operation (Figure 3). Alloys based on bismuth in combinations with antimony, tellurium or selenium are referred to as low temperature materials and can be used at temperatures up to around 450K.

The intermediate temperature range – up to around 850K is the regime of materials based on alloys of lead while thermoelements employed at the highest temperatures are fabricated from silicon germanium alloys and operate up to 1300K. Although the above mentioned materials remain the cornerstone for commercial/practical applications in thermoelectric generation, significant advances have been made in synthesising new materials and fabricating material structures with improved thermoelectric performance. Efforts have focused primarily on improving the material's figure-of-merit,[5,6] Typical of these are the filled skutterudites[7] and the clathrates[8] A material which is a promising candidate to fill the temperature range in the ZT spectrum between those based on bismuth telluride and lead telluride is the semiconductor compound β-Zn_4Sb_3[9]. This material possesses an exceptionally low thermal conductivity and exhibits a maximum ZT of 1.3 at a temperature of 670K.. During the past decade material scientists have demonstrated that low dimensional structures such as quantum wells (materials which are so thin as to be essentially of two dimensions, 2D), quantum wires (extremely small cross section and considered to be of one dimension, 1D, and referred to as nano-wires), quantum dots which are quantum confined in all directions and superlattices (a multiple layered structure of quantum wells) provide a route for achieving significantly improved thermoelectric figures-of-merit[10] The reduced dimensions of these structures result in an increase in phonon interface scattering, a consequent reduction in lattice thermal conductivity and resulting improvement in Z. Achieving comparable thermal conductivity reductions in bulk material has proved challenging Early success in reducing the thermal conductivity through use of fine grained compacted material was invariably accompanied by degradation in electrical properties[11-13.] Recently however   it has been reported that this unwanted increase in electrical resistivity has been   minimized in nanoparticle compacted material resulting in an improved figure of meri[14] . A result which opens up new frontiers.

THERMOELEMENTS AND MODULES

In practical applications   a module   consists in essence, of a large number of thermocouples connected electrically in series and thermally in parallel to form the multicouple structure shown schematically in figure 4 This is the basic building block of a thermoelectric generator which is shown schematically in figure 5 . Heat from one of a variety of sources is supplied to one side of the module, the hot side, and rejected at a lower temperature from the other side, the cold side. Semiconductors employed in thermoelectrics are by their nature poor conductors of heat. Consequently very efficient thermal insulation has been developed to ensure that as much as possible of the heat from the hot source passes through the thermoelements.

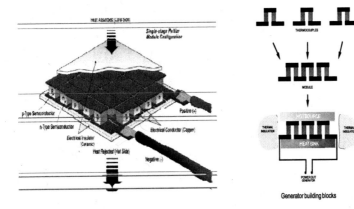

Figure 4 multicouple module   Figure 5 Generator building blocks

The essential components of a generating system are the heat source, thermoelectric convertor(modules) and   appropriate heat exchangers. For efficient operation it is essential that as much as possible of the available heat passes through the convertor. Consequently, the latter should interface effectively and efficiently in the collection of heat at the hot side and rejection of heat at the cold side. It is evident from figur 3 that thermoelectric materials operate at their maximum figure –of-merit over a very narrow range of temperature whilst a generator operates over as wide a temperature range as the thermoelectric material allows .Functionally grading and segmenting the thermoelements are two procedures which overcome

Functially graded and segmented thermocouples

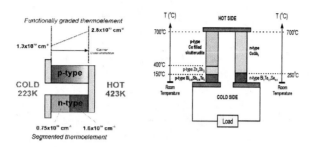

Figure 6 High performance couple   Figure 7. Segmented

this limitation to some extent. The temperature at which the figure- of- merit maximises can be shifted by altering the carrier concentration along the elements length (functionally grading) While in segmentation, materials whose figure- of- merit maximise at the temperature of operation, are butted together .

Silicide/Bi-Te Cascaded Modules
Komatsu Ltd.

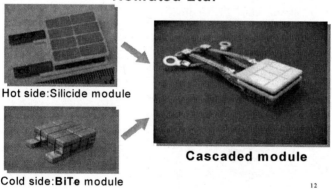

Hot side:Silicide module

Cold side:**BiTe** module

Cascaded module

Figure 8 examples of cascaded modules (Komatsu)

Figure 8 is an example of a high performance module comprised of functionally graded /segmented thermoelements mounted in cascade The thermoelements of the Bi₂Te₃ module was fabricated by joining two Bi₂Te₃ based material with figure-of-merit maxima at 273K and 383K. The maximum figure efficiency of the thermocouple over the range 23-423K was estimated to be 10% compared with the 8.8% for standard Bi_2Te_3 based material [15]

Ring-Stack Thermoelectric Module

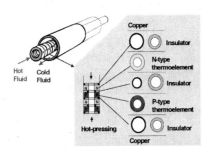

Figure 9 Schematic ring module   Figure 10 Prototype module

When heat flows in a radial directions, attaching plate-shaped modules around a cylindrical heat source is often complicated. An appropriate configuration for the case of radial heat transfer should be cylindrical geometry. In Figures 9 and 10 are shown a tube-shape module. It is a coaxial arrangement of a large number of flat annular "washers" with alternate 'washers' comprised of n-type and p-type thermoelectric material and connected electrically in series. They are joined alternately at their at their inner and outer peripheries by interleaved copper rings in the manner of compressed concertina bellows. The gaps between n- and p-type "washers" filled with electrically and thermally insulating materials.

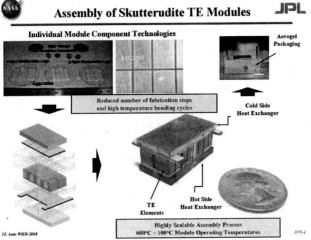

Figure 11 The assembly of a high performance module (J.P. Fleurial, JPL)

At the forefront of generator research are attempts to develop high performance multicouple module arrays that can readily be scaled up to a 100kW system for space power application. Fgure 11 is a flow diagram of the assembly for an eight couple module using skutterudite materials comprised of two parallel arrays of four couples in series. Capable of operating at temperature of up to 975K The challenge is to determine the best bonding methods and to minimising the number of high temperature assembly steps ..

Figure 12 Double helix liquid to liquid heat exchanger

Crucial to efficient generator performance is the avaiblity of high efficiency heat exchangers at both the hot and cold sides of the generator Currently designs operate at around 50%efficiency . It figure 12 is show na double helix spiral channel machined in a copper block . The design effectively increases the area of cooling or heating fluid in contact with the generator modules.

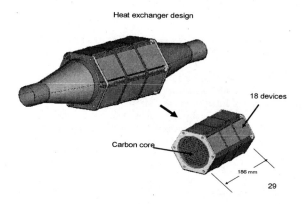

Figure 13 Porous carbon block hot side gas heat exchanger (Calsonic –Kansei Ltd)

Extracting the maximum amount of heat available in gases, as in vehicle exhaust waste heat recovery, presents different problems . Figure 13 is a schematic of a heat exchanger developed by Calsonic –Kansei to extract heat from vehicle exhaust gases. The prototype is comprised of a carbon block matrix which operates at an efficiency of around 50%..Careful design of aperture configurations is required in order to prevent build-up of an unacceptable engine back pressure..

GENERATORS

Power ranges of thermoelectric generators (TEGs), sources of heat together with typical applications are collected in Figure 14. Thermoelectric generators range over 15 orders of magnitude from the milli-microwatt gas-miniature thermocouple arrays, integrated into a semiconductor chip to the multi-hundred watt isotopic powered generators deployed in space exploration. Generator applications can conveniently be categorised by their source of heat into fossil fuel, isotopic and waste heat powered.

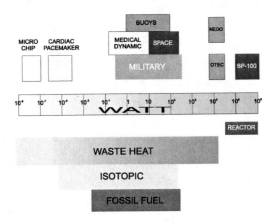

Figure 14 Thermoelectric generator applications, power ranges and sources of heat.

FOSSIL FUELLED

These application cover generators which utilise fossil fuel as a heat source . Typical fuels are propane, kerosene, coal and wood. The first fossil fuel powered generators are generally attributed to the efforts in the former Soviet Union of Ioffe et al in the late 1940's, when thermoelectric generators were used to power radios in remote regions. Probably the most publicised example is shown in figure 15[16].This is also the first example of a parasitic application where thermoelectrics produces electrical power when the device is primarily used to provide light. More modern generators have enjoyed modest but steady growth in applications and it is estimated that during the past 20 years more than 12000 fossil fuelled generators have been commissioned world–wide in industrial/commercial applications.[17]

The most frequent use of fossil fuelled generators has been in the communications industry; radio, television, microwave and telephone, where the modularity and

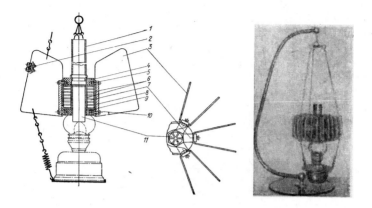

Figure 15 Early example of waste heat thermoelectric generation (L Anatychuk)

autonomous operation of a thermoelectric generator are clear advantages over land lines, engine driven generators, or solar-cell alternatives.

TEGs can also provide cathodic protection at power levels of several hundred watts and in remote regions are widely employed for this purpose. When used in the protection of pipelines transporting natural gas the supply of essentially free fuel enables them to compete economically with the more established sources of electrical power.

Figure 16A   Figure 16B (L.Anatychuk)

Figure 16A shows a gas fuel 20 W power generator for automation in gas-distributing stations. 3B shows a generator for the same purpose, where Joule-Thomson effect due to gas pressure reduction is used instead of gas combustion.

Although efficient utilisation of available fossil fuel presents problems in burner design, multi-fuelled burners have been developed in the USA for use with locally available fuel sources. Such generators which range in power output from 120 to1500 watts have found military applications where their multifuel capability and silent operation in tactical situations prove desirable features.[18]

Figure 17 Radial geometry module

In figure 17 is shown a generator comprised of two radial geometry modules[19] The thermoelectric elements in the module are made from PbSnTe doped to have either p or n-type semiconductor properties. One thermoelectric module has 325 couples with each couple consisting of a p-type element and an n-type element. The two modules were arranged in tandem. The lower one has a flat heat transfer surface at the hot side while the upper one contains a number of heat-conducting fins to increase convective heat transfer from the flue gases to the thermoelectric module (The total power output of the integrated system reached 1017.1W.

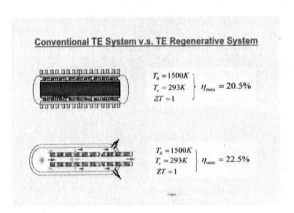

Figure 18 Schematic conventional and regenerative thermoelectric systems

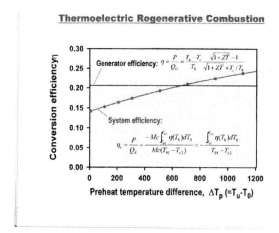

Figure 19 Generator performance comparison

The conversion efficiency of a thermoelectric system can be increased by preheating the gaseous fuel before combustion as shown schematically in figure 18 [20]

In figure 19 is displayed the effect on generator efficiency of preheating temperature difference. Evidently from the figure preheating the combustants results in an increase of 10% in the system's thermoelectric conversion efficiency.

ISOTOPIC

The concept of employing isotopic energy as a source of heat resulted in a proliferation in the application of thermoelectric generators. The long life of the heat source coupled with its high energy density facilities the production of

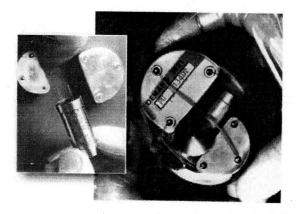

Figure 20 Isotopic powered cardiac pacemaker battery. (UKAEA)

thermoelectric generators that can operate unattended over long periods. Initially developed to provide power in space vehicles, radioisotope powered generators (RTG's) have found application in a variety of terrestrial, marine and medical applications.Thermoelectric power generation has enjoyed its greatest success in space applications.[21] The development of Systems for Nuclear Auxiliary Power (SNAP) was begun in the United States in 1955. The first application of radioisotope power in space accompanied the successful launching of a generator in June 1961. The thermoelectric power of 4 watts generated was a fraction of the satellites requirement and during the first decade of space exploration solar photovoltaic supplied the primary electrical power to earth orbiting space crafts such as Explorer and numerous other space vehicles in programmes such as Ranger, Surveyor, Luner Orbiter, Pioneer and Mariner.

GENERAL PURPOSE HEAT SOURCE
RADIOISOTOPE THERMOELECTRIC GENERATOR

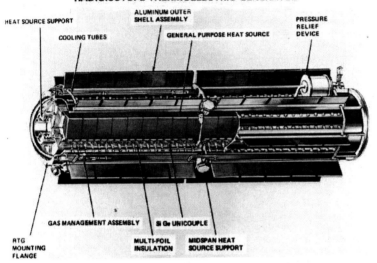

Figure 21 The GPHS-RTG (US DoE)

However in the 1970's attention turned towards the distant planets and in deep space missions to Jupiter and beyond TRG power offers distinct weight advantages over solar cells. Between 1961 and 2006 the United States has deployed 41 RTG's and one nuclear reactor powered generator providing power for 25 space systems. Yhe General purpose heat sourse- radioisotopic thermoelectric generator is the current main stay onboard power source, Shown in figure 21 it weighs 54.5Kg, measures 1.13m long and 0.42m in diameter. At the beginning of life it delivers 290W(e) at 28 volts

WASTE HEAT
Vast quantities of waste heat are discharged into the earth's environment much of it at temperatures which are too low to recover using conventional electrical power generators. Waste heat powered

thermoelectric generators can be discussed in terms of their power levels.

Low power is less than a watt. The first miniature thermoelectric generator fabricated using integrated circuit technology is shown in Figure 22[22].Alternate n- and p-type active thermoelements are ion implanted into an undoped substrate. Several hundred thermocouples are connected electrically in series and occupy an area approximately 25mm square.

The generator which was designed specifically to provide sufficient electrical energy to power an electronic gas flow monitoring system, in excess of 1.5 volts, could be supplied when a temperature difference of a few tens of degrees was established across the device. Any available heat source such as the surface of a hot water pipe would provide sufficient heat flux to provide the required voltage. utilising waste human body heat to power a thermoelectric watch battery.[23]

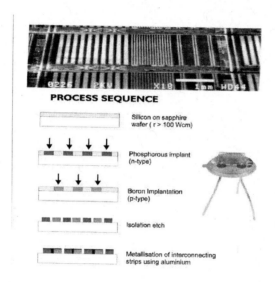

Figure 22 Miniature thermoelectric generator

The first multiwatt generators using waste warm water as a heat source is shown in figure Identified as a. Watt-100 mark 11 it incorporated improved bismuth telluride modules and generates 100 watts at a power density approaching 80kW/m3. The system was scalable enabling 1.5kW to be generated[24] A building block 100W generator powered, was exhibited at the 3rd World Energy Conference at Kyoto in 1998.

Figure 23   Watt-100 thermoelectric generator

Vehicle exhaust gases provide an attractive source of heat as shown in figure 24 with exiting potential for thermoelectric generation Typical available heat depends upon engine speed and temperature and for a family size car is around around 70kW at 560C Evidently if this heat could be extracted with 100% efficiency then, even with available thermoelectric materials several kilowatts of electrical power could be produced .

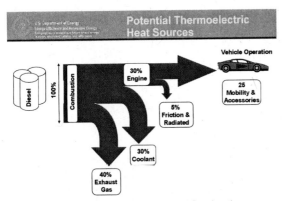

*Spark Assisted Gasoline internal Combustion
Engine (Light Truck or Passenger Vehicle)*

Fogure 24 Typical Energy inventory from Diesel powered vehicle (US DoE)

Both General Motor and BMW are pushing ahead with the use of thermoelectrics to improve automobile fuel economy[25]. BMW has moved their schedule up to have a thermoelectric generator providing a minimum of 10% improvement in fuel economy in their model year 2010 series 5. General Motors is on the same time line to introduce their generators. In figure 25 is shown a thermoelectric generator attached to the underneath of a BMW.
The Japanese are also progressing on target with a national programme to develop by 2010

thermoelectric generators employing high performance materials, for a wide range of domestic and commercial uses. These include automobile and heavy machinery application.

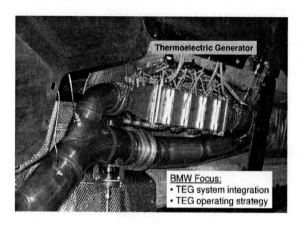

Figure 25 A thermoelectric generator fitted on the underside of BMW. (US DoE)

The first demonstration in Europe of a family sized Volkswagen car fitted with a thermoelectric generator was unveiled at the Thermoelektrik-Eine Chance Fur Die Atomobillindustrie meeting held in Berlin,October (2008). Under motorway driving conditions a 600W(e) output is claimed. The additional electrical power serves to meet around 30% of the car's electrical requirement. This reduces the engine's mechanical load such as that due to the alternator and results in a reduction in fuel consumption

Thermoelectric waste heat recovery operating at an achievable 10% efficiency offers a tremendous potential, both in reducing the impact of the increasing use of fossil fuels and the discharge of environmentally harmful gases into the atmosphere. According to General Motors a 10% fuel economy resulting from the use of thermoelectric waste heat recovery equates in their 2010 production to a 707,000 tons reduction in CO_2 for regular vehicles and 550,000 tons for hybrids with a saving of 100 million gallon of fuel..

CONCLUSIONS

A principle goal of thermoelectricians is to make available thermoelectric generators that can be employed over a wide range of environmentally friendly applications A system thermal to electrical conversion efficiency of around 10-12% is generally though as the bench mark at which thermoelectric technology becomes economically viable. Achieving this target will require an almost doubling of the present material figure-of-merit of bulk material to around 2. together with an improvement in heat exchange efficiency from the present 50% to around 65%. Recent reports that the enhanced phonon scattering observed in thin film nanostructured films has been obtained in nanoparticle size bulk materials makes the material target achievable Current material programmes to develop and fabricate high performance scalable modules employing skutterudite materials for space applications are also on track. Substantial progress has also been made in developing generators for exhaust waste heat recovery employing Bi_2Te3 technology. However, .extending the generator's operating temperature will require considerable efforts to successfully utilise higher temperature. thermoelectric materials such as those based on lead telluride . Other factors such as toxicity, weight, material cost and availably will need to be addressed before achieving globalisation of thermoelectric generation and over a wide spectrum of applications.

REFERENCES
[1] D.M.Rowe and C.M.Bhandari Modern Thermoelectrics, Holt Technology, ISBN 0-03-910433-8 (1983)
[2] A.F.Ioffe, Semiconductor Thermoelements and Thermoelectric Cooling, Infosearch, London, (1957).
[3] I..B. Cadoff and E. Miller, (Eds.) Thermoelectric Materials and Devices, Reinhold, New York, (1959).
[4] D.M.Rowe,(Ed.) CRC Handbook of Thermoelectrics, Section D, CRC Press. (1994)
[5] C.M. Bhandari, and D.M. Rowe, Thermal Conduction in Semiconductors, Wiley Eastern Ltd., 1988.
[6] G.A. Slack, Design concepts for improved thermoelectric materials, Mat.. Res . Symp,. Proc.. Vol 478, 47-54,(1997).
[7] J.P. Fleurial, T. Caillat.and A. Borshchevsky,. Skutterudites- an Update. Pro. ICT 97. p1 (1970)..
[8] G.S. Nolas, Semiconductor Clathrates; A PGEC System with potential for thermoelectric applications. Thermoelectric Materials Mat. Res. Symp. Proc. Vol. 545, 435-442, (1999)..
9 T.Cailalt, J.P.Fleurial, and A Borschevsky, Preparation and thermoelectric properties of semiconducting Zn_4Sb_3. J.Phys. Chem. Solid, 58, .1119, (1997)
[10] M S Dresselhaus and J P Heremans Recent developments inlLow dimensional thermoelectric materials in Thermoelecrics Handbook, Macro to Nano.. Ed D M Rowe,.CRC Press (2005)
[11] D.M.Rowe, Carrier transport properties of sintered germanium-silicon alloys, PhD thesis, University of Wales Institute of Science and Technology, (1968).
[12] C.M. Bhandari and D.M. Rowe, Fine grained silicon germanium alloys as superior thermoelectric materials' Proceedings 2nd International Conference on Thermoelectric Energy Conversion, University of Arlington, Texas, USA, March, .32-35,(1978).
[13] D.M. Rowe, V.S. Shukla and N. Savvides 'Phonon scattering at grain boundaries in heavily doped fine grained silicon-germanium alloys', Nature, Vol.290 No.5806 April, .765-766.(1981).
[14] B Poudel et al.. HigBeh-Thermoelectric Performance of Nanostructured Bismuth Antimony Telluride Bulk Alloys , Science express, published on line 20 March 2008:10.1126/science 1156446 .
[15] V.L.Kuznetsov.,L.A. Kuznetsova A.E.Kaliazin D.M. Rowe,High performance functionally graded and segmented BI_2Te_3 based materials For thermoelectric power generation, *Journal of Materials Science* **37**, No. 14,

2893-2890,(2002)

[16] L Anatychuk-private communication

[17] W.C. Hall, 'Terrestrial applications of thermoelectric generators' Chapter 40 . 503-513, CRC Handbook on Thermoelectrics, Ed. D.M.Rowe, ISBN 0-8493-0146-7 (1994).

[18] G. Guazzoni, Thermoelectric generators for military applications, Proc. 4th International Conference on Thermoelectric Energy Conversion, University Of Texas at Arlington, . 1-6, (1982.)

[19] K. Qiu, A.C.S. Hayden A 1 kW thermoelecrc power generation system for micro –cogeneration. Proceedings 6th European conference on thermoelectrics , Paris, France, July 2008

[20] F.J. Weinberg, D.M. Rowe and G. Min, Novel high performance small-scale thermoelectric power generation employing regenerative combustion systems', J. Phys. D: Appl. Phys. 35 , L61-L63. ISBN S0022-3727(02)31732-7, 2002

[21] D.M Rowe. 'United States Thermoelectric Activities in Space', Proceedings VIIIth Int. Conf.on Thermoelectric Energy Conversion, 10-13 July 1989, Nancy, France, pp. 133-142.

[22] D.M. Rowe, 'Miniature Thermoelectric Converters', British Patent No. 87 14698, (1988)

[23] S. Saiki, S.I. Takeda, Y. Onuma. and M. Kobayashi, Thermoelectric properties of deposited semiconductor films and their application, Electrical Engineering in Japan, Vol. 105,2, . 387, (198)

[24] D.M.Rowe,The NEDO/Cardiff Thermoelectric project to economically recover low temperature waste heat, 17th Int. Conf on Thermoelectrics, Nagoya, Japan, .18-24 ISSN 1094 2734, (1998).

[25] J Fairbanks, Thermoelectric applications in vehicles-status 2008. proc.6th European conference on thermoelectics, Paris, France, July (2008).

Chemical Thermodynamic in Thermoelectric Materials

Jean Claude Tedenac, R.M. Marin-Ayral, Didier Ravot, F. Gascoin

Université Montpellier 2 LPMC-UMR-CNRS-5617, Montpellier,France

Abstract.

High ZT thermoelectric materials are obtained by a good knowledge of the materials involved in their fabrication. The phase transformations and phase stabilities of new materials are still unknown; consequently, a thermodynamic study of these systems is needed. The development of thermodynamic and kinetic databases of such practical materials is important for the microstructural evolution of the materials during processing and service for improving the knowledge.

In this paper we will present some peculiarities of this materials at the light of a phase diagram analysis and show finally how the Calphad method is necessary to study such multicomponent materials and to determine the best synthesis process (single crystals growth, powder metallurgy,...).

Exemples of advanced thermoelectric materials taken is this paper are: $Pb_{1-x}Sn_xTe$ Zn_4Sb_3 and $Co_{4-x}Ni_xSb_{12}$. They are studied from a thermodynamic point of view and we will show the relationships between calculated phase diagram, microstructure constitution and processes.

Introduction.

The efficiency of a solid-state thermoelectric engine primarily depends on the chemical compositions of materials and then of phase stabilities. The improvement of the figure of merit in thermoelectric materials depends on multiple factors regarding the microstructures and the physical properties (1).

Thermoelectric materials are typically multicomponent systems and a thermodynamic approach is necessary to study such systems (2).

The processes involved in the fabrication of a thermoelectric component lead to other problems:

a -Solidification processes cause segregations in single phases and, depending on the composition of the melt, eutectic precipitations.

b - Temperature dependence of the material composition causes point defects leading to imperfect electronic properties.

c - Hot pressing processes cause element diffusion in the grains and at the boundaries leading to non-equilibrium material with properties changing with time and temperature.

Information's on phase transformation, thermodynamic stabilities and process modelling of the new materials presently studied for thermoelectricity are scarce.

In this paper we show how to use a thermodynamic database in a general Calphad approach (3,4) and to understand phases stabilities of thermoelectric materials. Material microstructures after sintering and thermal behavior are checked with the isothermal section at high temperature. Moreover, Gulliver-Scheil simulations (5,6) (infinite diffusion in liquid phase and no diffusion in solid phase) can be used to determine the solid fractions in a solidification process and the final microstructure. After an example on lead telluride material we will present the features of a general Calphad approach in thermoelectric antimonides.

1.   Using a thermodynamic database for the synthesis of a ternary material: $Pb_{1-x}Sn_xTe$.

The two intermetallic compounds PbTe and SnTe are narrow band-gap semi-conductors and display complete solubility in the whole composition range. The carriers concentration in PbTe materials changes with the chemical composition (p-type in the metal rich side and n-type in the tellurium rich side). A thermodynamic database has been contructed by H.L. Lukaet als (8). The free energy function description of the system has been modelled in order to describe the departure from stoichiometry at all temperatures. This description allowed us to calculated the ternary phase diagram by the Calphad procedure and apply such data to control ball milling process.

Usually ternary alloys based on 20-30 mol% of tin are studied for thermogenerator materials (9). Some important problems in the constitution of materials can be solved using the corresponding phase diagram. The definition of the homogeneity range of PbSnTe materials in the whole range of temperature can easily be made by a thermodynamic description of the section. The thermoelectric material of p type presenting the higher ZT is close to the composition $Pb_{0.8}Sn_{0.2}Te$. It can be described by a cut (isopleth section) of the ternary lead-tin-tellurium at 20 at% of tin.

The section is presented in figures 1 a and b. In the central part of the phase diagram the homogeneity zone of the thermoelectric material is shown. This description is very useful for determining precisely the conditions of quenching and the synthesis of the single phased material as well as to check a ball milling process.

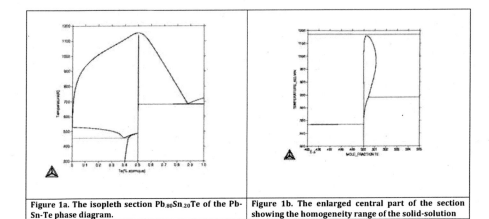

Figure 1a. The isopleth section Pb.80Sn.20Te of the Pb-Sn-Te phase diagram.	Figure 1b. The enlarged central part of the section showing the homogeneity range of the solid-solution

2. The Calphad technique.

The CALPHAD method has been widely developed for metallic systems, for oxide systems and semi-conductor systems as, for example, gallium arsenide, cadmium telluride and lead telluride systems (8,10,11)

It is very important to introduce the point defects in the modeling into the phases in order to map the defects as a function of chemical composition. Due to its difficulty, this characteristic, important for physical applications, was not taken into account in previous assessment of semi-conductor systems. In the last decade, the defect modelling in semi-conductors has been introduced and used (example in the Cd-Te system, ref. 8).

Calculation of phase equilibria in a multicomponent system is obtained by the minimization of the total Gibbs energy, where G is a summation of the Gibbs energy of all phases that take part in each equilibrium as is expressed by equation (1):

$$G = \sum_{i=1}^{p} n_i G_i^0 = \min.(1)$$

The thermodynamic description of the whole system requires the assignment of thermodynamic functions to each phase. The main interest of the CALPHAD method is the use of a variety of models to describe the free-energy functions of the various phases of the system as a function of temperature, pressure and composition. Then the Gibbs energy function of a phase j can be written as in equation (2):

$$G_m = \sum_i x_i \left(G_0^i - {}^i G_0^{phys} \right) + G_m^{phys} - T S_m^{ideal} + {}^E G_m \ (2)$$

The effects of particular physical phenomena (as magnetism) are taken into account by subtracting them from the description of the end-members (reference state) and it is introduced for the solution through the contribution G^{phys}_m. The last term on the right hand side is the excess term.

Usually the temperature dependence of the Gibbs energy is expressed as a power series of T where a, b, c and dn are coefficients and n are integers (3). The pure elements are represented by the same function according to the database of A. Dinsdale (10).

$$G = a + b.T + c.T.\ln(T) + \sum d_n.T^n \ (3)$$

By taking into account all the parameters the Gibbs energy of phases are then represented by three contributions as expressed in equation (4).

$$G^0 = {}^{ref}G^0 + G^{ideal} + G^{xs} \quad (4)$$

The first term of the right-hand side corresponds to the Gibbs energy of a mechanical mixture of the components, the second one corresponds to the entropy of mixing for an ideal solution and the third term, the so-called excess term, represents all the deviations from ideality.

Thermodynamic modeling of phases is the core of the CALPHAD approach. Most systems have few strictly stoichiometric compounds, so it is important to model phases deviating from the ideal stoichiometry. The most used model for solution phases are random substitutional, or ordered sublattices. In ordered solid phases, the Wagner-Shottky model is used for describing small deviations from stoichiometry (which are non-interacting defects). Intermetallic binary compounds are generally non-stoichiometric and can exist in a large range of composition. In this case the most used model is the sublattice model (3). Moreover, additional informations from physical properties (chemical potentiels of electrons and holes measured by transport experiment) can be put inside this model and implement the database.

All the models can be used for ternary and multicomponent systems by adding high order interactions terms in the expression of the excess free energy.

In such high ordered systems the Gibbs energy must be calculated from extrapolation of the excess quantities of constituent sub-systems. Several methods can be used and among them the geometrical Muggianu method is the most used (3). Consequently the Gibbs energy of a n-component solution phase must be determined by the n-1 energies using the method given by Figure 3.

In fact relationship between chemical potentials of the phases and the system gives non linear equations that can be solved by numerical methods such as Newton-Raphson methods. The CALPHAD type packages use mathematical methods to minimize the Gibbs functions.

$$G = \sum x_i G_i^0 + RT \, x_i \ln x_i + G^{ex}$$

BINARY   Assessment G_{bin}^{ex}

⇓

TERNARY   Extrapolation G_{bin}^{ex} + Assessment G_{ter}^{ex}

⇓

QUATERNARY   Extrapolation $(G_{bin}^{ex} + G_{ter}^{ex})$ + Assessment G_{qua}^{ex}

⇓

HIGH ORDER   Etc...

Figure 2 The extrapolation method used in the Gibbs energy calculation.

Experimental data used for these calculations are those concerning the phase diagram measurements, enthalpy of formation of the compounds, Cp for CoSb and chemical potentials measured by electrochemistry. According to the description adopted in the SGTE datadase,(12), the Gibbs energy of pure elements is taken with reference to the enthalpy of the elements in the SER state (standard element reference), the elements in their stable state at 10^5 Pa and 298.15 K.

2. The zinc-antimony binary system.

The next generation of thermoelectric materials will be built up with intermetallic phases. Many of these materials are based on antimony systems such as Sb-Zn or Co-Sb and they will be multicomponent in order to optimize thermoelectric properties. Over this last decade number of research on Zn-Sb based materials have been made on synthesis and physical properties and large ZT values have been obtained (14,13). In these studies one problem is missing: optimization of the fabrication processes related to the understanding of phase relationships.

The problem occuring in zinc antimony materials is due to difficulties in the crystal growth of materials. Having initially a high figure of merit, those T.E. materials could have better properties with some achievement in the metallurgical processes (15). These difficulties are easily understood by a complicated phase diagram. Up to our recent paper (16) the different phase equilibria determinations lead to six different versions of this phase diagram. These differences were due to real difficulties in the phase stabilities and consequently to the phase equilibria determination.

Using a Calphad analysis we have determine the phase relationship in this binary system and explained the behaviour of the material. The figures 3a and 3b present the calculated phase diagram obtained by this way (17). All the important information for crystal growth (temperatures, compositions, phase stabilities, liquidus shape,...) are indicated in this diagram.

Moreover, by a crystal structure analysis of the compounds informations about the sub-lattice modeling which can be used. Associating phase diagram and crystal chemistry, the two intermetallic compounds can be modelled by a four sub-lattice model according to the references 3 and 4, the model for the Gibbs energy calculations should be a four sublattice model with two antimony sublattices (Sb$_1$, Sb$_2$), one zinc and one vacancy sublattice including Zinc as interstitial. These calculations are presently in progress.

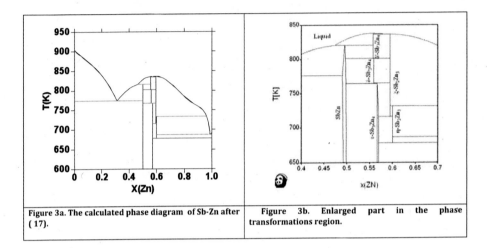

Figure 3a. The calculated phase diagram of Sb-Zn after (17).

Figure 3b. Enlarged part in the phase transformations region.

4. The ternary Co-Ni-Sb.

The optimized skutterudite thermoelectric material belongs to the multicomponent system: Ce-Co-Fe-Ni-Sb . The purpose of our research is to build up a database with five elements in order to improve the materials behaviour. In this aim, it is necessary to assess the binaries and ternaries as reported in a general CALPHAD procedure (section 1).

One key system for studying these materials where the skutterudite phase should exist with a certain homogeneity range is the Co-Ni-Sb ternary. First it is necessary to assess the border binaries (Co-Sb, Ni-Sb).

Most of the binary systems having a skutterudite type compound were experimentally studied but no thermodynamic assessment was available until now. The skutterudite phase (cobalt triantimonide) is a stable compound, but the nickel materials appear unstable, the extension of solid solution have to be studied in the ternary.

First, modeling the Co-Sb system has been done (19,20). In this system three intermetallic compounds are formed: CoSb, CoSb$_2$ and CoSb$_3$. In Nickel-Antimony only two compounds are formed: CoSb, and CoSb$_2$; consequently, in this assessment the regular solution model was adopted for the liquid and terminal solution solids. Concerning the intermetallic compounds the no experimental information on arange homogeneity was given in the literature, so the Gibbs energy of CoSb$_2$, CoSb$_3$ are stoichiometrically modeled and described by a sub-regular model. The 1:1 compound was described with a three sublattice model in order to take into account the departure from stoichiometry on the both sides of the solid solution. The modeled phase diagrams are presented in Figure 4a. The figure 4b show the shape of the Gibbs energy functions at 500℃ for the binary Co-Sb, a proove of the relevance of the models.

The liquidus curve and the invariant temperatures are known to good accuracy. This result will be an aid for understanding the thermodynamic behaviour of the ternary and quaternary phases. It has been possible to calculate the temperature of the metastable melting point of CoSb$_3$ (1285K instead of 1210K for the peritectic decomposition of the 1:2 phase). One can observe that these temperatures are not so far and it could explain the problems occuring in the fabrication of the materials as well as the lack of 1:3 compounds for iron and nickel systems.

The most useful result is the shape of the phase diagram around CoSb$_3$. In the binary the liquidus curve and the invariant temperatures are known to good accuracy.

As it was explained in section 1, modelling a ternary system need to know the binaries. In that case one can observe that a phase is dominant (the 1:1 nickel arsenide phase) and a good modelling of the ternary and realated binaries depends on the model used in this phase. The choice of the model for the Gibbs energy calculations should be a three sublattice model with two metal sublattices (Co₁ , Co₂) including vacancies, and one antimony sublattices (Sb).

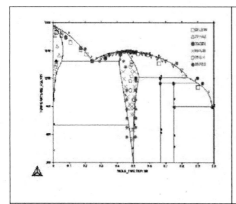

Figure 4a. The calculated Co-Sb from ref.18. The lines represent the result of calculations and the dots represent the different experimental data.	Figure 4b. The Gibbs eneregy curves of the dirfferent phases of the system at 500°C.

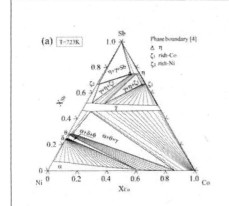

Figure . Calculated isothermal section at 500°C showing the phase relationships between the skutterudite phase(η), (γ) and antimony rom ref.20.	Figure . Calculated isothermal section at 600°C showing the phase relationships between the skutterudite phase(η), (γ) and antimony. The dots represent the experimental results.

Chemical Thermodynamic in Thermoelectric Materials

Conclusions

Using the CALPHAD method we have described phase diagrams of some thermoelectric materials and show the possibilities of this kind of study for the fabrication processes. We have modelled the Sb-Zn, Co-Sb phase diagrams which are the first step of the constitution of our database. The ternary Co-Fe-Sb is experimentally determined and the modelling is presently in progress.

References

1. Fleurial J.P.,Caillat T.,Borshevsky A., Morelli D.T. and Meisner, Proc. 15 th Intl Conf. on Thermoelectrics, Piscataway, NJ, (1996), p 9.
2. Tedenac J.C., Marin-Ayral R.M., Ravot D., Record M.C.
   Proc. XX Int. Conf. On Thermoelectrics, ICT'01, (2001) 371-374
3. Saunders N, Miodownick P. "CALPHAD, a comprehensive guide", Pergamon (Exeter, 1998), p 261.
4. Kattner U., J.O.M.,49 (12), (1997), 14-19.
5. Scheil E. Z. fur Metallkunde Vol.34, (1942), pp 70
6. Gulliver G.H. Metallic alloys, Verlag griffin,Vol.25, (1922), pp. 120-157.
8. Kattner U., Lukas H.L., Petsow G., Gather B., Irle E., Blachnik R. Z. fur MetallK.Vol. 79, (1988), pp 32-40.
9. Bouad N., Ph. D Thesis Montpellier (France) 2001
10. Ansara I., Chatillon C., Lukas H L., Nishizawa T., Ohtani H., Ishida K., Hillert M., Sundman B., Argent B. B., Watson A., Chart T. G., Anderson T.. Calphad, 18-2, (1994), 177-222.
11. Chen Q., Hillert M.., Sundman B., Oates W.A., Fries-Gomes S.G., Schmid-Fetzer R., J of Elec. Mat.,Vol 27 No 8, (1998), pp 961-971.
12. Dinsdale A. Calphad, Vol.15, (1991), pp 317-425.
13. Sales B.C., Mandrus D., Williams R.K. Science 272, (1996), p 1325.
14. Fleurial J.P.,Caillat T.,Borshevsky A., Morelli D.T. and Meisner, Proc. 15 th Intl Conf. on Thermoelectrics, Piscataway, NJ, (1996), p 9.
15. Izard V. ., Ph. D Thesis Montpellier (France) 2001
16. Izard V., Record M.C., Tedenac J.C., G. Fries S. , Calphad, Volume 25, Issue 4, December 2001, Pages 567-581
17. Jing-Bo Li, Record M.C., Tedenac J.C., Journal of Alloys and Compounds, Volume 438, Issues 1-2, 12 July 2007, 171-177.
18. J.C. Tedenac, M.C. Record and V. Izard, Mat. Res. Soc. Symp. Proc. 691 (2002), pp. 23–28.
19. Yubi Zhang, Changrong Li, Zhenmin Du, Tai Geng, Calphad, Volume 32, Issue 1, March 2008, Pages 56-63.
20. Yubi Zhang, Changrong Li, Zhenmin Du, Cuiping Guo, J.C. Tedenac, Calphad, 2009,on line.

STRENGTH OF BISMUTH TELLURIDE

A. A. Wereszczak, T. P. Kirkland, O. M. Jadaan,* and H. Wang**

Ceramic Science and Technology
Oak Ridge National Laboratory
Oak Ridge, TN 37831

*   College of Engineering, Mathematics, and Science
    University of Wisconsin-Platteville
    Platteville, WI 53818

** Diffraction and Thermophysical Properties Group
    Oak Ridge National Laboratory
    Oak Ridge, TN 37831

ABSTRACT

The fast-fracture or inert strength of a commercially available p- and n-type bismuth telluride was measured in uniaxial and biaxial flexure at 25 and 225°C. The processing method is recognized to produce material anisotropy, so the orientation effect on strength was also explored. Two-parameter Weibull strength distributions for p- and n-type, orientation, and temperature are contrasted. N- and p-type had equivalent strength in the RZ-orientation but the n-type was slightly stronger in the RR-orientation. The strength in the RZ-orientation was approximately twice that of the RR-orientation for both the n- and p-types. Strength decreased by approximately 15% between 25 and 225°C. Lastly, as-machined and cut surfaces produced equivalent strengths with the latter showing less strength scatter.

INTRODUCTION

The use of thermoelectric materials (TMs) and devices (TDs) for heating, cooling, and power generation at modest maximum temperatures (e.g., 200°C) has occurred for decades now. Further developments of compositions usable to 500°C are actively underway at many companies, universities, and federal research laboratories because achieving that capability would enable the exploitation of many additional waste-heat-recovery power-producing situations.

Kingery's [1] thermal resistance parameter, R_{Therm}, is useful to refer to in context with TMs,

$$R_{Therm} = \frac{S_{Tens}(1 - \nu)\kappa}{\alpha E} \quad ,$$

(1)

where S_{Tens} is tensile strength, ν is Poisson's ratio, κ is thermal conductivity, α is the coefficient of thermal expansion or CTE, and E is elastic modulus. For improved thermomechanical resistance against effects caused by thermal gradients or thermal transients, one desires R_{Therm} to be as large as possible. The parameters ν, κ, α, and E are materials properties and are essentially unchangeable for any given TM under consideration. For TMs, the minimization of κ is purposely and primarily sought because that achievement improves thermoelectric efficiency. Additionally, all TMs the authors have characterized typically have large a large CTE (> 10 ppm/°C). The ν for most TMs does not vary far

131

from 0.25. The E's for TMs typically range between 50-125 GPa; however, the E is virtually unchangeable within a given class of TMs (e.g., skutterudites, TAGS, tellurides, etc.). Therefore, the intent to make the R_{Therm} for TMs as large as possible is primarily hindered by the inherently low κ and typically high CTE.

The remaining parameter in Eq. 1 is therefore strength. The S_{Tens} for brittle materials, including TMs, is not a material property in a strict sense. The S_{Tens} for brittle materials is a function of many parameters (e.g., size) that are either intrinsic or extrinsic to the material. Some of those effects are shown in this study. Because of the brittleness (i.e., low fracture toughness), the S_{Tens} is anticipated to be at least one order of magnitude lower than compressive strength in TMs as is the case for polycrystalline ceramics; therefore, for conservative design, testing will and should focus on the measurement of a tensile strength in TMs. Of all the parameters used in the right-hand side of Eq. 1, the manufacturer of TMs and TDs can only tangibly increase R_{Therm} by increasing S_{Tens}. Thus, the valid measurement of strength in TMs, the identification of those flaws that limit strength, the active reduction of those flaw sizes to increase S_{Tens}, and appropriate strength-size-scaling of TMs in TD design all need to be executed to ultimately achieve the highest probability of survival in a TM and TD in service.

An example of a TD and a TM leg are schematically shown in Fig. 1. Thermoelectric materials and TDs function with a thermal gradient existing across them. This thermal gradient creates a stress gradient owing to the TM's coefficient of thermal expansion. Tensile stresses result on the edges (1-dimensional or 1D) and surfaces (2D), and within the volume (3D). For proper and confident design and reliability prediction of TDs containing these TMs, the tensile strength of all three must be known and managed.

Because of their geometry, the strength of TMs could be limited by three different types of strength-limiting flaws as shown in Fig. 2. Though fractographical results are not presented in this study, a brief discussion of flaw types is warranted owing to the later reporting of strengths. Volume-type (or bulk-type) flaws usually are intrinsic to the material and indicative of how well it was processed. Surface- and edge-type flaws usually are extrinsic to the material and are indicative of either how well the material was treated or used after it was processed. Examples of each of the three types are listed in Fig. 2. Additionally, "hybrid flaws" can result from the combination of two flaw types and limit strength as well. Examples of the states of an edge, surface, and volume of a bismuth telluride (Bi_2Te_3) bend bar are shown in Fig. 3. The S_{Tens} can be limited by any of these states. The critical action for S_{Tens} for 1D, 2D, and 3D states is to understand the relationship between maximum tensile stress and flaw size, and work to reduce the latter. The complexity of S_{Tens} illustrates why it is not a material property in a strict sense.

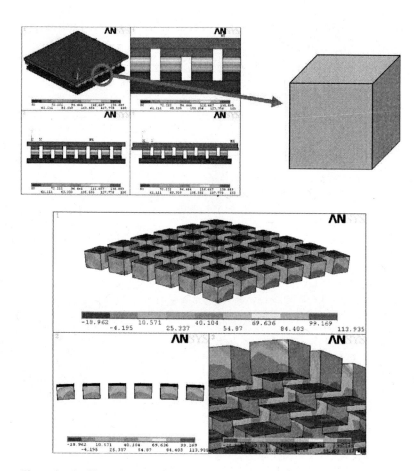

Figure 1. An illustration of a thermal gradient across the width of a thermoelectric module and the basic prismatic shape of a thermoelectric leg used in it, and the resulting First Principal stresses in the legs that result.

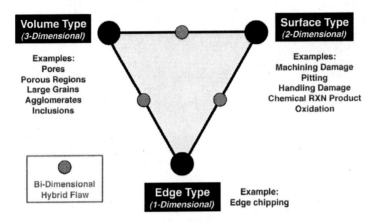

Figure 2. A triangle may be used to illustrate the strength-limiting flaw classifications for brittle materials. Because thermoelectric legs have edges, all three of the above types can be operative.

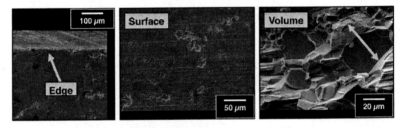

Figure 3. Examples of 1D (edge), 2D (surface), and 3D (volume) type flaws that could be operative in thermoelectric legs. The arrow in the latter shows a relatively large grain.

Lastly, for TD design optimization and reliability maximization, one ultimately needs to take the S_{Tens} distribution for each of those three cases (defined as "censored strength distributions") measured with simple test coupons, and concurrently size-scale them based on the stress gradient of the test coupon and final component. Much of this method can be examined in ASTM C1239 [2]. That concurrent adaptation usually tells the user which is the most problematic and which they should turn their attention to (e.g., minimize 3D or 2D or 1D flaw size) to improve the overall mechanical reliability. This too is not featured in this study but the inclusion of its discussion is meant to show the reader how (censored) data could be used to design TDs using the censored S_{Tens} data. The Weibull strength size scaling function for 1D, 2D, and 3D distributions are shown in Eqs. 2-4 respectively.

$$P_f = 1 - \exp\left[-k_L L \left(\frac{S}{\sigma_0}\right)^{m_L}\right] \qquad (2)$$

$$P_f = 1 - \exp\left[-k_A A \left(\frac{S}{\sigma_0}\right)^{m_A}\right] \qquad (3)$$

$$P_f = -\left[-k_V V \left(\frac{S}{\sigma_V}\right)^{m_V}\right] \qquad (4)$$

where P_f are probabilities of failure, $k_L L$, $k_A A$, and $k_V V$ are effective lengths, areas, and volumes, respectively, S is applied tensile stress, σ_{OL}, σ_{OA}, and σ_{OV} are length, area, and volume scaling parameters, respectively, and m_L, m_A, and m_V are Weibull modulus for length, area, and volume distributions, respectively. Such analysis is done numerically with life prediction software such as CARES/Life (Connecticut Reserve Technologies, Inc., Gates Mills, OH), and an example of such analysis can be found in Ref. [3].

The primary goal of this study was to establish a strength database for a reference TM, chosen here to be Bi$_2$Te$_3$, that would be usable for future comparisons to new, higher-TMperature-capable TMs. As part of that, strength specimens were chosen so as to promote fracture initiated from different type of flaw. Additionally, parametric strength comparisons of n- and p-type Bi$_2$Te$_3$, orientation, temperature, and how the specimens were machined or sliced into shape were sought.

EXPERIMENTAL PROCEDURE

Bismuth telluride specimens were acquired from a commercial manufacturer (Marlow Industries, Dallas, TX)[1]. The manner in which these materials (both n- and p-type) are fabricated produces transverse isotropy, so prismatic- and plate-shaped test coupons that enable uniaxial and biaxial flexure strength testings were sought for both orientations (designed as "RR" and "RZ") as illustrated in Fig. 4. The uniaxial (3-pt-bend) and biaxial flexure test fixtures used are shown in Fig. 5. Testing was done both at room temperature and 225°C to examine any changes in strength as a function of temperature. To produce all the specimens, plates were first ground parallel and some were sliced into prismatic bars. For those bars, this produced two different surface conditions designated as "as-machined" and "cut". Any differences in strength caused by surface-located flaws produced by those two surface preparation methods were examined. Lastly, one surface of a set of the plates was metallographically polished to examine what effect the machining had on strength. At least 15 specimens were tested in all sets.

A universal test machine (Instron, Canton, MA) was used for all the strength testing. Owing to the small specimen sizes, the specimen preparation methods and procedures described in ASTM C1161 [4] and ASTM C1499 [5] for uniaxial and biaxial flexure strength, respectively, required adaptation to calculate strength from the maximum failure load. Commercial statistical software (WeibPar, Connecticut Reserve Technologies, Cleveland, OH) was used to fit the strengths to a two-parameter

---

[1] Commercial materials and equipment are identified in this paper to adequately specify the experimental procedure; however, this should not be construed to imply endorsement by the authors, their institutions, or the organizations supporting this work.

Weibull distribution using maximum likelihood estimation. 95% confidence ratio rings were determined for each set using WeibPar and were used to assess the various independent parameter's effects on strength.

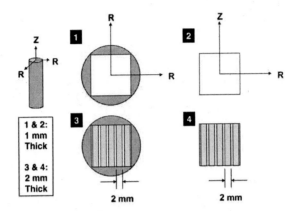

Figure 4. Four combinations of specimen and orientations were studied. Plates for biaxial (1 & 2) and bend bars for uniaxial flexure testing captured the RR- and RZ-orientations (3 & 4).

Figure 5. Specimen geometries and test fixturing.

RESULTS AND DISCUSSION

The characteristic strength of the RZ-orientation was approximately twice that of the RR-orientation for both the n- and p-types (80 vs. 40-50 MPa) as shown in Fig. 6. The strengths of the n- and p-type for the RZ-orientation were equivalent, whereas the n-type was slightly stronger in the RR-orientation.

Obviously, it will improve mechanical reliability for the TD manufacturers to take advantage of the stronger orientation.

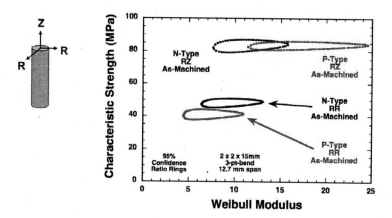

Figure 6. Three-point bend strength as a function of type and orientation.

The characteristic strength for both types and both orientations decreased by approximately 15% between 25 and 225°C. This is shown in Fig. 7. When a leg is subjected to a thermal gradient, tensile stresses primarily develop where the coolest temperatures are and are compressive where the warmer temperatures exist. Therefore, this (tensile) strength decrease at 225°C perhaps would not be problematic in service.

The bend strengths from specimens whose tensile surfaces were either as-machined or cut were equivalent as shown in Fig. 8. There was a trend that the as-machined surfaces exhibited more scatter in their strength distribution as evidenced by the lower Weibull modulus.

Lastly, polishing the Bi_2Te_3 produced a flexure strength approximately 10% greater than that from the as-machined surfaces for both orientations. This is illustrated in Fig. 9. The polished strength may be viewed as a "strength potential" of the material. This result shows that the as-machined surface preparation methods used are not significantly detrimental to strength.

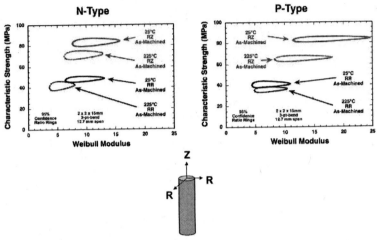

Figure 7. The strength decreased by approximately 15% between 25-225°C.

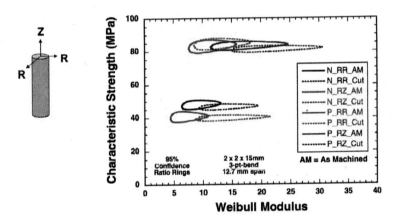

Figure 8. Three-point bend strengths as a function of type, orientation, and surface condition.

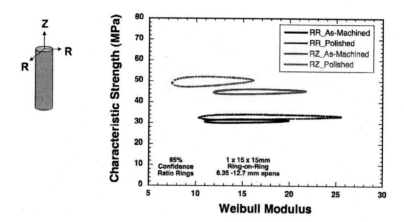

Figure 9. Ring-on-ring biaxial flexure strength as a function of as-machined and polished surface conditions and orientation at 25°C for the p-type Bi₂Te₃.

CONCLUSIONS

Several trends in bismuth telluride (Bi_2Te_3) strength were observed. N- and p-type had equivalent characteristic strengths in the RZ-orientation but the n-type was slightly stronger in the RR-orientation. The strength in the RZ-orientation was approximately twice that of the RR-orientation for both the n- and p-types. Strength decreased by approximately 15% between 25 and 225°C. Lastly, as-machined and cut surfaces produced equivalent strengths with the latter showing less strength scatter. These results provide reference strengths that can be compared to strengths of developmental thermoelectric materials.

ACKNOWLEDGEMENTS

Research sponsored by the U.S. Department of Energy, Assistant Secretary for Energy Efficiency and Renewable Energy, Office of Vehicle Technologies, as part of the Propulsion Materials Program, under contract DE-AC05-00OR22725 with UT-Battelle, LLC. The authors thank ORNL's M. Lance and H. Wang for reviewing the manuscript, and H. -T. Lin for assisting with the scanning electron microscopy

REFERENCES

[1]   W. D. Kingery, "Factors Affecting Thermal Stress Resistance of Ceramic Materials," *Journal of the American Ceramic Society*, 38:3-15 (1955).
[2]   "Standard Practice for Reporting Uniaxial Strength Data and Estimating Weibull Distribution Parameters for Advanced Ceramics," ASTM C 1239, Vol. 15.01, ASTM, West Conshohocken, PA, 2008.
[3]   O. M. Jadaan and A. A. Wereszczak, "Probabilistic Design Optimization and Reliability Assessment of High Temperature Thermoelectric Devices," *Ceramic Engineering and Science Proceedings*, [3], 29:157-172 (2008).
[4]   "Standard Test Method for Flexural Strength of Advanced Ceramics at Ambient Temperature," ASTM C 1161, Vol. 15.01, ASTM, West Conshohocken, PA, 2008.
[5]   "Standard Test Method for Monotonic Equibiaxial Flexure Strength of Advanced Ceramics at Ambient Temperatures," ASTM C 1499, Vol. 15.01, ASTM, West Conshohocken, PA, 2008.

Materials for
Solid State Lighting

SULFOSELENIDE PHOSPHORS AND NANOPHOSPHORS FOR SOLID-STATE LIGHTING

H. Menkara, T. R. Morris II, R. A. Gilstrap Jr, B. K. Wagner, and C. J. Summers
PhosphorTech Corporation*
351 Thornton Rd, Suite 130, Lithia Springs, Georgia 30122

ABSTRACT

We report a comparative study between the properties of sulfoselenide materials and conventional LED phosphors. These properties include high refractive index and unique thermal and optical behavior that are observed in both bulk and nanocrystal materials. Solid forms of these luminescent materials were developed for the first time, offering real potential for index-matching to InGaN without the use of epoxy or silicone at the interface. We also report on doped ZnSe nanocrystals where the absorption properties are a function of shell thickness and the emission properties are a function of the activator. Both the solid (macrophosphors) and the nanocrystalline (nanophosphors) forms of these sulfoselenide materials offer unique potential for achieving high outcoupling and conversion efficiencies in UV/blue LEDs by reducing both scattering and total internal reflection losses.

Keywords: Solid-state lighting, phosphors, nanophosphors, nanocrystals, white LEDs.

INTRODUCTION

Solid-state lamps (SSLs) and light emitting diodes (LEDs) are among the most efficient converters of electrical energy into light. Additionally, they have the advantages of long lifetime, excellent reliability, DC power operation, light weight, small size, and excellent resistance to mechanical shock and vibration. Although there has been a steady improvement in conventional lighting (i.e. incandescent, halogen, and fluorescent lamps) over the past 20 years, the efficiency of LEDs has improved to the point where they are replacing incandescent and halogen lamps in traditional monochrome lighting applications, such as traffic lights and automotive taillights. Recent breakthroughs in the efficiency of blue InGaN diodes have resulted in phosphor-coated LEDs becoming a serious contender for conventional incandescent white lighting. While commercial white LEDs have demonstrated commercial efficiencies between 30-70 lm/W, luminous efficiencies as high as 130 lm/W have been reported recently in the laboratory.[i]

High-brightness LEDs are expected to operate at high optical power densities, with junction temperatures reaching ~200° C, a wide range of external temperatures, and humidity levels that could exceed 90% relative humidity (RH). Therefore, it is of critical importance to investigate the operational limits of any LED phosphor and to develop techniques and materials that will improve the phosphor and LED lamp performance under extreme operating conditions.

OBJECTIVES

In this work, we report new approaches to synthesize bulk, solid form, and nanocrystalline phosphor materials that have the potential to achieve high quantum and extraction efficiencies compared to traditional phosphors. By improving the phosphor and LED extraction efficiencies as well as their thermal and chemical stability, the potential for high-brightness and high luminous efficiency white LEDs will be significantly enhanced. The goal is to provide the LED designer with a wide range of efficient and stable multi-color phosphor options that make it possible to build high performance LEDs with a large number of color spectra for virtually any indoor/outdoor application.

LIGHT EXTRACTION

The total optical conversion efficiency in a phosphor-based diode lamp sequentially depends upon the following factors: (1) efficient optical coupling of the LED excitation into the phosphor; (2) efficient absorption of this excitation energy and its transfer to the activator; (3) efficient radiative emission, of appropriate wavelength, from the phosphor; and (4) efficient out coupling from the phosphor to air with minimal scattering. The first and last effects depend on external factors determined by the optical properties of InGaN and the luminescent materials. The optical coupling efficiency between the LED and phosphor is determined by the difference in their relative refractive indices (n). If the difference is large, significant losses occur due to Fresnel reflections and total internal reflection (TIR). TIR occurs when light is incident from a high refractive index medium into a lower refractive index medium at an angle greater than the critical angle (which is a function of the ratio of the refractive indices). Additional losses can occur due to scattering by the micron-sized phosphor particles. This is the case for blue/UV emission from an InGaN LED device (n ~ 2.7) propagating into a lower index surrounding having particulate properties (e.g. phosphor dispersed in silicone, epoxy, etc.). Index matching between the LED and the phosphor materials will entirely eliminate TIR reflections, light piping, and Fresnel reflections; however, traditional micron-size phosphor particles will be subject to scattering losses. Even if a bulk phosphor with high refractive index is used, encapsulating such a phosphor using low-index polymers (silicone, epoxy, etc.) eliminates its high-index advantage and results in increased scattering and reduced light extraction from both the diode and the phosphor.

Figure 1 illustrates the role that index-matching can play in improving the extraction efficiency of an LED/phosphor system. Both theoretical (Figure 1) and experimental[ii] data show that even a very simple (single phosphor layer) index-matched device can produce over 50% increase in efficiency (with no polymer encapsulation), compared to a conventional white LED structure. The Cu-doped ZnSeS (yellow-orange) phosphor system has a refractive index around 2.5, closely matched to that of InGaN, while the refractive index of the standard YAG phosphor is only 1.8. However, when a standard (n~1.38-1.4) polymer is added to the equation, the difference in extraction efficiencies between the ZnSeS and YAG are reduced to less than 10%. Due to the lack of available high index (2.4-2.6) polymer encapsulants, maximizing the LED extraction efficiency using a polymer+phosphor (particles) system is not possible due to the low refractive index at the InGaN die/polymer+phosphor interface. A more effective approach would be the use of a high index solid luminescent phosphor structure that can be intimately coupled to the InGaN chip without the conventional silicone/epoxy polymer encapsulant, as illustrated in Figure 2.

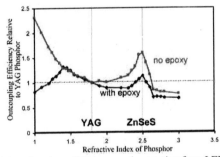

Figure 1. Theoretical data showing ~50% increased extraction from LED with index-matching when no epoxy encapsulant is used.

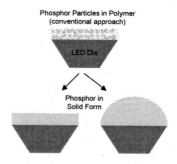

Figure 2. Conventional phosphor-based LED approach compared with solid phosphor approach.

CONVENTIONAL BULK MICROPHOSPHORS

Conventional bulk powder phosphors, currently used in LED devices, have enabled a wide-range of lighting and display products, which would not have been possible or economical with LEDs alone. The most popular YAG:Ce LED lamp offers high efficiency (>90%) and good thermal and chemical stability, but with limited color flexibility and a very broad emission spectrum extending into the near-infrared. Therefore, other phosphor systems have recently been investigated and produced as potential alternatives or supplements to the YAG-based phosphors used in most of today's white LEDs. The most common alternative LED phosphors are based on the following material systems:

1. Eu-doped orthosilicates: $(Ba,Sr,Ca)_2SiO_4$:Eu
2. Eu-doped alkaline-earth silicon nitrides, $(Sr,Ca)_2Si_5N_8$:Eu and oxynitrides: (SiONs): $(Sr,Ca)Si_2O_2N_2$:Eu
3. Eu-doped chlorosilicates $Ca_{(8-x)}Eu_x(Mg,Mn)(SiO_4)_4Cl_2$
4. Eu-doped thiogallates and thiogallates-selenides: $(Sr,Ca,Ba)Ga_x(S,Se)_y$:Eu
5. Cu-doped zinc sulfoselenides: $ZnSe_xS_{(1-x)}$:Cu

None of the above phosphors offer a complete solution for all LED pump configurations or color applications and each has its advantages and disadvantages. Limitations of conventional phosphors include low refractive index and micron-sized particles resulting in increased light scattering and reflection losses. Furthermore, high thermal quenching results in reduced phosphor conversion efficiency in high power LEDs. These "micro-phosphors", currently being used in virtually all white LED products, are fabricated and used as random-shaped particles ranging in size from 5 mm to 25 mm, depending on the material. Regardless of their chemical composition or intrinsic efficiencies, the particulate nature of conventional phosphors results in significant scattering of light, whether it originates from the LED or the phosphor itself. The severity of the light scattering depends mainly on the phosphor's scattering length and refractive index, as shown in Figure 3. Therefore, even if a high refractive index (2.6-2.8) phosphor is used, the light outcoupling efficiency can still be significantly reduced as a result of light scattering from the individual particles. In order to achieve high outcoupling efficiency, a phosphor material with both a high refractive index and large mean scattering length is needed. The sulfoselenide material systems provide for the possibility of achieving both of these goals simultaneously using either a "macrophosphor" or a "nanophosphor" structure, as will be shown in the following sections.

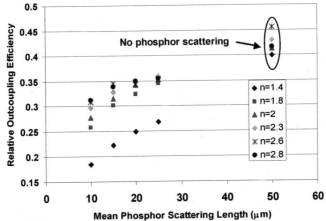

Figure 3. Theoretical data showing effect of phosphor refractive index
and scattering length on outcoupling efficiency.

SOLID FORM MACROPHOSPHORS

A novel macrophosphor synthesis technique was developed recently in which it was possible to achieve Cu/Ag doping and efficient green to red luminescence from solid forms of ZnS/ZnSe materials, as shown in Figure 4. These solid form of ZnSeS:Cu,Ag phosphors varied in size from 1 to several millimeters, depending on the shape and size of the original precursors. These were obtained by high temperature sintering of yellow-orange ZnSeS:Cu,Ag phosphors and ZnSe/ZnS crystals then slowly cooling to room temperature. Unlike powder phosphor systems, macro-sized crystalline phosphors provide the unique advantage of high density and lower scattering compared to micron-sized particles. In addition, if these crystals can be directly fused onto LED dies without any low-index polymers, it will be possible to take full advantage of the ZnSeS high refractive (InGaN-matching) index properties to maximize light extraction.

Figure 4. Solid form of ZnSeS:Cu,Ag phosphors with bright luminescence from green to red.

In order to demonstrate the potential of these materials, the luminescent performance of the solid form ZnSeS (SFZSS) yellow phosphor crystals were investigated by fabricating a white LED using a blue (~460 nm) device, as shown in Figure 5. The spectral performance of the white LED shown below was compared to YAG:Ce powder phosphor and to our standard ZnSeS:Cu,Ag (ZSS) powder phosphor, when pumped by a blue LED operating at ~455 nm. The luminescence and colorimetric results from these samples are shown in Figure 6 where the SFZSS phosphor appears to have more favorable properties to YAG:Ce compared to powder ZSS phosphor. For example, the peak emission of SFZSS is at 577 nm with a FWHM of ~99 nm, and exhibits properties much closer to YAG (565-570 nm peak and a FWHM~121) nm than the ZSS phosphor with peak luminance at 600 nm and a FWHM~89 nm. When combined with a blue LED, the SFZSS phosphor achieved color chromaticity cordinates CIE (x=0.332, y=0.317) and color temperature CCT=5530 compared to CIE (x=0.318, y=0.226) and CCT=9475 for the ZSS phosphor. On the other hand, the YAG:Ce with a blue LED has CIE (x=0.316, y=0.302) and CCT=6595.

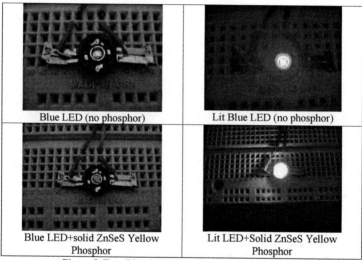

Figure 5. Top: Blue LED. Bottom: White LED using blue LED and a solid ZnSeS:Cu,Ag yellow crystal.

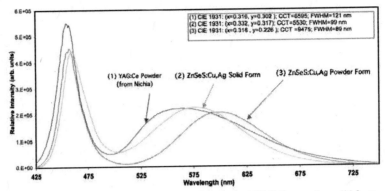

Figure 6. Spectra of blue LED combined with standard YAG:Ce powder, solid form
ZnSeS:Cu,Ag and powder form ZnSeS:CuAg.

NANOCRYSTALLINE (NANO) PHOSPHORS

As previously discussed, the performance of white LEDs using conventional phosphors suffers from several limitations due to the intrinsic properties of most phosphor materials. The fixed absorption and emission character of conventional phosphors is an intrinsic property that has posed challenges to a variety of display and lighting applications. When used to create white LEDs, conventional micron-sized phosphor tends to scatter light which reduces the LED's overall efficiency. The size of these particles also results in non-homogeneous light output from each lamp.

Within the last 5 – 10 years, the potential incorporation of quantum dots to alleviate the problems of color rendering and efficiency in white LEDs has received increasing attention.[iii] Quantum dots (QDs) are semiconductor nanoparticles that offer the greatest degree of emission color tuning via manipulation of the intrinsic band gap.[iv] One of the most efficient systems of this type is CdSe which can be tuned to emit from red to blue.[v] Because its absorption band also spans the visible range with increasing particle size, this system can efficiently absorb blue light and emit at virtually any visible wavelength desired simply by controlling its size. Typical particle sizes of less than 10nm have the added benefit of eliminating internal light scattering as they are much smaller than the wavelength of light which they produce. Additionally, homogeneous solutions are very easy to attain as these materials can be suspended easily in a variety of solvents. The inherent toxicity of Cd, raises significant doubt as to its widespread adoption for LED technology. However, this quantum dot system and several others have already found use as biological tagging agents for cells and proteins.[vi] In addition to the Cd problem, the application of undoped QDs to solid-state lighting is fundamentally problematic due to two primary issues: self-absorption and thermal quenching. Self-absorption is due to the overlap between emission and absorption bands, which results in a portion of the emission being re-absorbed by adjacent (larger) particles. This effect causes specific problems in multi-color systems because short wavelength emissions are significantly absorbed.[vii]

A more difficult problem to address is that of the thermal quenching of luminescence. Rather than recombining radiatively across the band gap, free excitons within a quantum dot are readily bound to lattice phonons at temperatures exceeding 100°C.[viii,ix,x] An example of this phonon-driven effect can be found in a Sandia National Lab study on the thermal stability of CdS and CdSe quantum dots encapsulated in both epoxy and silicon as is routinely done for application to a LED device.[xi] In that study, the luminescence from CdS quantum dots decreased by 50% at 75C. Similar thermal quenching problems occur in standard phosphor materials, and the luminescence efficiency of even some of the most thermally stable phosphors can drop by more than 10% at 150C. In a typical high-brightness LED, junction temperatures can reach as high as 200C, so the thermal luminescence quenching of both conventional phosphors and undoped quantum dots is expected to be even more severe.

An effective approach to solving many of these issues that plague conventional phosphors and QDs is by doping the nanocrystalline particles with impurity atoms that introduce a recombination center within the band gap. The absorption character of this "nanophosphor" system will be unaffected because excitation is still achieved by promotion of electrons in the valence band to excited states in the conduction band. The emission character however, will be red-shifted to longer wavelength as excitons recombine on the impurity levels within the band gap. Just as in the case of micron-sized doped semiconductors, the precise value of this shift (termed the Stoke's Shift) is a function of the type of transition involved and impurity atoms utilized. In this way, a system of different sized doped nanophosphors may have no overlap between emission and absorption bands, as shown in the Mn-doped ZnSe nanophosphor system in Figure 7. Additionally, there is recent evidence suggesting that a significant enhancement in resistance to thermal quenching can be achieved by the doping of QDs.[xii,xiii] It is theorized that excited states in doped QDs will preferentially recombine at doping sites rather than lattice phonons or surface states when these systems are heated above room temperature.

Successful synthesis of nano-crystalline phosphor materials was conducted using an automated system consisting of an electronic dispenser and PC-controlled flask heating mantle. This system was used for continuous unattended growth of Mn-doped ZnSe nanocrystals. This was achieved by first dissolving manganese stearate and then purging/filling with nitrogen and degassing at 110°C for 1 hour under Argon. Selenium powders were then dissolved at 140°C under nitrogen and injected into the manganese stearate flask. Finally, zinc acetate was dissolved at 140°C under nitrogen and then slowly injected into the heated flask over an extended period of time. An automated electronic dispenser system was programmed to periodically introduce small quantities of precursors in order to allow for a homogenous growth and doping of the nanocrystals.

Emission and absorption measurements were conducted on the synthesized ZnSe:Mn nanocrystals at room temperature. A strong yellow emission at ~585nm (from Mn doping) and a weaker band gap emission from ZnSe were observed, as shown in Figure 7. In order to tune the band gap of this system closer to blue diode pump wavelengths of 440-460nm, the ZnSe shell thickness was increased. Due to the larger band gap of the MnSe cores, excited state excitons are confined to the ZnSe shell region. Accordingly, a thicker shell reduces the quantum confinement effect and narrows the band gap, which has a reciprocal relationship to wavelength. The challenge in this task was maintaining the balance needed between Se, Zn, and the surfactant octadecylamine (ODA) concentrations at all times during the reaction. As the optimum Se:Mn ratio was reached, additional Se injections were required to maintain a Se-rich solution during subsequent Zn injections. Sufficient ODA was needed at all times to prevent particle agglomeration. The ZnSe shell material was therefore added in three steps with intermittent injection of additional Se + ODA solution. Figure 8 presents absorption spectra from a series of aliquots taken directly from the reaction solution at the completion of each ZnSe shell application. Great care was taken to ensure the concentration of particles in the solvent solution was the same for each sample.

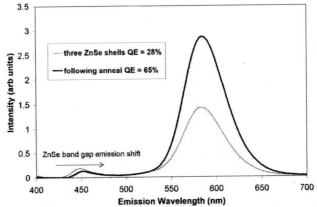

Figure 7. Photoluminescent emission spectrum of ZnSe:Mn nanocrystals prepared by the balanced multilayered synthesis techniques described above.

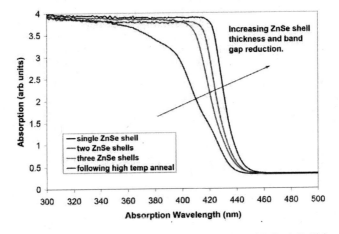

Figure 8. Absorption spectra of ZnSe:Mn QDs as a function of ZnSe shell thickness.

Two key features can be observed from the above spectra. First, the sample with a single shell of ZnSe maintains the "wavy features" associated with quantized conduction band states. This feature disappears as additional ZnSe shell layers were heterogeneously attached and is due to a natural widening of the size distribution. Second, the high-temperature annealing of the sample with three ZnSe shells produced a significant shift to longer wavelengths. We believe that this effect is likely a factor of enhanced crystallinity within the ZnSe layer resulting from higher temperature diffusion of defects within the lattice. As each shell is applied, defects are expected to be generated at each ZnSe-to-ZnSe layer. Annealing should assist in passivating and removing these features making the three separate layers act as one complete zone. The effect on the emission intensity of the annealed samples is quite significant as shownin Figure 7. We can now utilize the information derived from the minor ZnSe band gap emission in the 440-450nm region. The emission from the annealed sample occurs at approximately 5nm longer wavelength which is entirely consistent with the difference observed in the absorption spectra. This is further indication that the band gap reduction and associated ability to be excited by longer wavelength light was achieved.

The morphology of this system is shown in the transmission electron microscopy (TEM) image of Figure 9. The average size of ZnSe:Mn particles is ~9nm with a size distribution range of about 2nm. Careful study of various TEM images revealed that addition of the third ZnSe layer has served to narrow the size distribution by preferentially attaching to the smaller particles in the solution. This is consistent with the effects of higher surface energy on an epitaxial growth phenomenon. The smaller particles in the system will have higher surface energy and therefore, attract ZnSe shell growth to a greater extent. We now believe that particle sizes of 15nm are possible with optimization of ZnSe growth conditions. This size will push the system absorption edge to the maximum theoretical value of the bulk band gap of ZnSe at approximately 460nm.

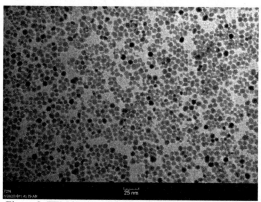

Figure 9. TEM image of ~10nm ZnSe:Mn nanophosphors.

CONCLUSIONS

The unique properties of various forms of sulfoselenide materials were investigated for solid-state lighting applications. Solid luminescent forms of these materials were developed and characterized for the first time, and have been shown to offer real potential for use as wavelength down-converters in LEDs. The nanocrystalline form of these materials was also synthesized and modified for increased blue light absorption by increasing the ZnSe shell thickness. Both forms of these materials offer the ability to simultaneously index-match to InGaN and provide reduced light scattering properties, which should contribute to enhanced light out-coupling efficiency in phosphor-based LEDs.

ACKNOWLEDGEMENT

This work was supported in part by funding from the U.S. Department of Energy (DOE) and the National Science Foundation (NSF). Any opinions, findings and conclusions or recommendations expressed in this material are those of the authors and do not necessarily reflect those of the sponsoring agencies

FOOTNOTES

* info@phosphortech.com; phone 1 770 745-5693; fax 1 770 828-0672; www.phosphortech.com

REFERENCES

[i] CREE Press Release, "Cree Demonstrates 131 Lumens per Watt White LED", www.cree.com/press/press_detail.asp?i= 1150834953712, June 2006.

[ii] Menkara, H., "Novel High Extraction Phosphors for White LEDs", Phosphor Global Summit, March 2005.

[iii] Hsueh-Shih Chen, Cheng-Kuo Hsu, and Hsin-Yen Hong, "InGaN-CdSe-ZnSe Quantum Dots White LEDs", IEEE Photon. Technol. Lett., vol. 18, no. 1, pp. 193-195, Jan. 2006.

[iv] V. Klimov, "Nanocrystal quantum dots. From fundamental photophysicsto multicolor lasing", Los Alamos Science, Nov. 28, pp.214-220, 2003.

[v] C.B. Murray, C.R.Kagan, and M.G.Bawendi, "Synthesis and characterization of monodisperse nanocrystals and close-packed nanocrystal assemblies", Annu. Rev. Mater Sci., 30, pp.545-610, (2000).

[vi] W. C. W. Chan, and S. Nie, Science, 281, p.2016, (1998).

[vii] J.P. Wilcoxon and P. Newcomer, "Optical properties of II-VI semiconductor nanoclusters for use as phosphors" SPIE Proceedings: Vol 4808 (2002).

[viii] Z. Y. Xu, Z. D. Lu, X. P. Yang, Z. L. Yan, B. Z. Zheng, J. Z., Xu, W. K. Ge, Y. Wang, and L. L. Chang, Phys. Rev. B 54, 11528 (1996).

[ix] Y. Tang, D. H. Rich, I. Mukhametzhanov, P. Chen, and A. Madhunkar, J. Appl. Phys. 84, 3342 (1998).

[x] S. Fafard, S. Raymond, G. Wang, R. Leon, D. Leonard, S. Charbonneau, J. L. Merz, P. M. Petroff, and J. E. Bowers, Surf. Sci. 361, 778 (1996).

[xi] Billie Abrams, Lauren Rohwer, Jess Wilcoxon, Stever Woessner, and Steve Thoma, "Photoluminescence Thermal Quenching of Encapsulated CdS and CdSe Quantum Dots for Solid State Lighting Applications", Abs. 1167, 204th Meeting, The Electrochemical Society, Inc., 2003.

[xii] Y. Hori, X. Biquard, E. Monroy, D. Jalabert, F. Enjalbert, Le Si Dang, M. Tanaka, O. Oda, and B. Daudin, Applied Physics Letters 84, No. 2, 206 (2004).

[xiii] R. A. Gilstrap Jr.a H. M. Menkara, B. K. Wagner, and C. J. Summers, "Doped Quantum Dots for Solid-State Lighting", Proceedings for the EL2008 Conference, September 2008, Rome, Italy.

Author Index